LAND, FOOD and PEOPLE

Based on the FAO/UNFPA/IIASA report
"Potential population-supporting capacities
of lands in the developing world"

FOOD AND AGRICULTURE ORGANIZATION OF THE UNITED NATIONS
Rome, 1984

The designations employed and the presentation of material in this publication do not imply the expression of any opinion whatsoever on the part of the Food and Agriculture Organization of the United Nations (FAO), the International Institute for Applied Systems Analysis (IIASA) or the United Nations Fund for Population Activities (UNFPA) concerning the legal status of any country, territory, city or area or of its authorities, or concerning the delimitation of its frontiers or boundaries.

P-04
ISBN 92-5-102079-5

Printed in Italy

Foreword

Man's demands for food, from the natural resources that sustain his very existence, have increased enormously in recent years. By the year 2000, a world population of more than 6 billion will require an agricultural output some 50 to 60 percent greater than in 1980. Demand for food and agricultural products in developing countries will double.

Are global land resources sufficient to achieve this increased production and feed future populations? Are country land resources sufficient to attain food self-sufficiency for our present and future populations? The first is a question of world-wide concern; and the second is a question increasingly being asked by countries seeking self-reliance.

Recognizing the importance of these issues, the Food and Agriculture Organization of the United Nations undertook in the 1960s, in cooperation with Unesco, a world-wide inventory of soil resources which culminated in the publication of the Soil Map of the World. The agro-climatic information base was concurrently developed by the Plant Production and Protection Division of FAO from data provided by National Meteorological Services through WMO and agro-ecological zones were determined for each major food crop. This enabled a first assessment of the physical potentials of land to produce food, compared with the food needs of future populations. The long-standing experience of FAO's Land and Water Development Division was employed to lead this interdisciplinary assessment covering no less than 117 developing countries, and I am pleased to record that the work was undertaken with the support of the United Nations Fund for Population Activities (UNFPA) in collaboration with the International Institute for Applied Systems Analysis (IIASA).

The results of this fine example of cooperation among countries and international institutions under the aegis of FAO, as contained in the Technical Report of the Project, were presented to the Twenty-second Session of the FAO Conference in November 1983, and clearly showed that:

— the potentials of land to produce food are limited and vary considerably between and within countries;
— there are considerable extents of "critical areas" where land resources are insufficient to provide the food needs of present/projected populations;
— more refined and detailed national studies, particularly in the identified "critical areas", are needed.

To enhance awareness of the findings of the study, it is considered appropriate to reproduce the results in a more "popular" format, which also explores the wider implications of the technical findings. The object of this present "popular" report is to achieve such public awareness, particularly with regard to the main conclusions, namely:

— by the end of the century, the entire potentially cultivable land of developing countries — almost three times the presently cultivated area — would barely be sufficient to feed expected populations if low level of input methods of cultivation were to be applied;
— increasing the levels of inputs, intensifying the use of land and expansion of cultivated area are all vital if future populations are to be even minimally fed. Aid, including food aid, will be indispensable in bridging the period until these measures can be implemented and a balance attained between land, food and population;
— the most critical phase of the population explosion and its most serious potential confrontation with land, water and energy in many areas, are perhaps still to come. The years from 1980 to 2020 will see the largest annual increase in population in history and one of the most disturbing features of this growth is that it will be fastest in those areas where land resources are least adequate to meet food needs.

While, because of the large number of assumptions made in the study, these conclusions are necessarily of an indicative nature, one need is overwhelmingly clear:

— every country should begin to plan now how it can balance its population with its potential for producing food from its own lands, in various combinations with the introduction of a higher level of inputs and management, investments in arresting land degradation and in land improvement work, while producing foreign exchange for necessary imports.

Our 1983 Conference has recognized the importance of such work and has recommended concentration of future activities at the country level with the collection of more detailed national data and incorporation of socio-economic and other issues including trade, forestry and fisheries potentials, environmental protection, financial resources and development of human resources.

It is hoped that the present report will accelerate the production of such national studies; within the limits of its financial resources, FAO is prepared to assist in such work.

Edouard Saouma
Director-General

Acknowledgements

This report was prepared by Paul Harrison, London-based development writer, on the basis of the 1983 FAO/UNFPA/IIASA Technical Report, "Potential Population Supporting Capacities of Lands in the Developing World," by G.M. Higgins, A.H. Kassam and L. Naiken, Food and Agriculture Organization of the United Nations (FAO), and G. Fischer and M.M. Shah, International Institute for Applied Systems Analysis (IIASA). The support of the United Nations Fund for Population Activities (UNFPA, New York) and the collaboration of the International Institute for Applied Systems Analysis (IIASA, Vienna) in the study are gratefully acknowledged.

Contents

Foreword iii

Executive summary ix

1. Origin and methods 1

2. The context of agriculture in developing countries 4

3. The findings 11

4. The present predicament 34

5. Outlook: potentials and pitfalls 46

6. What can be done? 56

References 78

Appendixes

1. Present and projected populations and ratios to present and projected population-supporting capacities by location 81
2. Statistical analyses of the 64 countries and territories critical for low inputs in the year 2000 89
3. Technical note: assessing input levels by yield 95

Figures

1.	Climate and soils	5
2.	The degradation threat	9
3.	The potential land base for rain-fed agriculture	10
4.	Populations of critical countries	13
5.	Present input levels by region	15
6.	The climatic imbalance	20
7.	Area and populations of critical zones	23
8.	The critical countries and territories, year 2000	25
9.	Land reserves and cultivated area	44
10.	Population growth, 1950-2000	48
11.	The potential for land reform	59
12.	Developing countries with low water availability	62
13.	Population: low and high projections, 2000-2100	71
14.	Expansion of irrigation	72
15.	Critical countries' attitudes to fertility	74

Tables

1.	Regional potential needs/ratios	16
2.	Population ratios	39
3.	Potential of actually cultivated land	41
4.	Progress on yields	54
5.	Future population ratios, low projection	68

Maps

1.	Potential population-supporting capacities, critical zones, 1975	97
2.	Soil and climate data base	99
3.	Critical countries, low inputs, 1975	101
4.	Environmental hazards	103

Executive summary

By the end of this century, the entire lands of developing countries — almost three times the present cultivated area — would barely be sufficient to feed their expected populations if traditional methods of farming continued to be used.

No less than 64 countries — 29 of them in Africa — would be unable to feed their projected population from their own land resources. Some 2 450 million hectares, almost two fifths of the land area, with 60 percent of the total population, would be carrying more people than they could support, representing a serious threat to human welfare and the environment.

These are the most challenging conclusions of the first attempt to answer scientifically the question: how many people are the lands of the developing countries capable of feeding?

The study carried out by FAO in collaboration with the United Nations Fund for Population Activities and the International Institute for Applied Systems Analysis reveals, however, that many of the 64 critical countries would be able to feed their expected populations from their own lands if they could raise the level of farming or slow down their rates of population growth. It also demonstrates a huge potential for food self-sufficiency in four out of five developing regions — and in the developing world as a whole.[1]

---

[1] The study herein referred to and upon which the present work was based is entitled *Potential population supporting capacities of lands in the developing world*, Technical Report of Project INT/75/P13, "Land resources for populations of the future." See Reference 3.

The methods and the limitations

The study used information from 117 developing countries (excluding those in East Asia) on soils and climates and compared these with the growth requirements of 15 major crops and grassland pasture to assess potential food production and the numbers of people this could support. Because the level of farming has a great influence on yields, the study was made for three levels of input:

low, using no fertilizers, pesticides or improved seeds and no long-term conservation measures — equivalent to subsistence farming;

intermediate, with some fertilizers, pesticides and improved seeds, conservation measures and improved cropping patterns on half the land;

high, with full use of all inputs, full conservation measures and the most productive mix of crops on all land — equivalent to Western European levels of farming.

Most African countries are currently using little more than the low level of inputs, while countries in South and Southeast Asia are on average just under half way toward the intermediate input level. The Near East is about two fifths of the way toward the intermediate level, and Latin America about one third of the way.

The project's basic results assume that *all* potentially cultivable land is used to grow nothing but staple food crops or pasture for livestock, and that the produce is equally divided among social groups to provide a basic calorie requirement. In reality, land would need to be set aside for forestry, vegetables, fibres, non-food cash crops, and so on, which would reduce the potential *production*. The average *consumption* per person would be increased by inequalities, rising incomes and changing age structures. To allow for these factors, the potential population that could be supported on all cultivable land would have to be reduced by at least one third.

The major findings - low inputs

• *In 1975, there were already 54 countries that could not support their existing population from their own lands using low inputs.* After deducting one third from the carrying capacity, and allowing for land actually cultivated, this number rises to 99 out of 117.

• *The potential overloading of certain zones poses a very serious ecological threat.* In 1975, around 2 450 million hectares — 38 percent of the entire land area — were carrying more people than they could sustain, on a long-term basis, with low inputs. No less than 1 165 million people lived in such zones, which are mainly cooler, or drier, or more hilly areas.

These areas often coincide with areas having a high risk of desertification, or areas where overcutting of trees for fuelwood is occurring.

• *Land degradation may depress food production in the long run by an average of 19 percent if no long-term conservation measures are taken.* The area of rain-fed cropland could shrink by as much as 544 million hectares — an area greater than the entire potential cropland of Southeast Asia. Thirty percent of Central America's rain-fed cropland could be lost — and 36 percent of Southeast Asia's.

• *By the year 2000, with population growth pushing ahead of the increase in potential production, the situation will have deteriorated for most countries: 64 countries — more than half the total — would then be critical if they used low inputs.* Indeed, 38 of these would be able to support less than half their projected populations.

• *The entire potentially cultivable lands of the 117 developing countries would be sufficient to support only 1.6 times the expected population of 2000* even if they were used only for food crops or grassland supporting livestock. This potential area is at least three times greater than the present cultivated area.

• *When one third is deducted for non-food crops, etc., only 7 percent more than the expected year 2000 population could be supported — and the population is expected to grow by another 43 percent by the year 2025.*

- *All five regions would be critical in 2000 on the land they are likely to be cultivating by then, using low inputs.* Southwest Asia, Central America, Africa and Latin America would be able to support around *half* their expected populations, while Southeast Asia would be able to support 85 percent.[2]

What must be done?

Over the next 40 to 50 years, the developing countries will see the largest additions to their populations in all history. On current UN medium projections, the year 2025 populations of the countries studied will total 5.1 thousand million — two and a half times larger than the 1975 figure of 1.96 thousand million.

The land resources study offers a first approximation assessment of the numbers each country can support from its own lands.

National surveys are needed to provide a more detailed picture and FAO can advise on assistance for these within the limits of its resources.

One need is overwhelmingly clear: *every country should begin now to plan how, over the next forty years, it can balance its population with its possibilities of producing food from its own lands or earning stable foreign exchange with which to purchase imports*. The more critical the country appears and the more rapid its present population growth rate, the more pressingly urgent is the need for long-term planning.

Country situations vary enormously, from those with vast land reserves to those with none, from oil exporters to low-income oil importers. Country policies will need to be equally diverse.

---

[2] Central America, in this study, covers also the Caribbean. See Appendix 1 for countries and territories included in each region.

Developing agriculture

Countries which may become critical can take a number of positive steps to increase agricultural production.

• Those with adequate land reserves can plan *expansion into their high reserve-potential areas*, with due attention to conservation of delicate ecosystems.

• For most countries, an *increase in the level of inputs* will be essential.

Raising inputs has a powerful effect on the numbers of countries likely to be critical. Of the 64 countries critical for low inputs in the year 2000:

• *28 countries and territories would cease to be critical if they could apply intermediate inputs*. These would be

in Asia: Bhutan, Viet Nam, Sri Lanka, and the Philippines;

in Central America: Guadeloupe, the Windward Islands, Trinidad and Tobago, Jamaica, the Bahamas, the Dominican Republic, and Guatemala;

in Africa: Tunisia, Morocco, Nigeria, Senegal, Sierra Leone, Benin, the Upper Volta, Togo, Mali, Ethiopia, Uganda, the Comoros, Malawi, Zimbabwe, Botswana, Swaziland, and Namibia.

• *A further 17 would cease to be critical if they could apply high inputs*. These would be

in Asia: Bangladesh;

in Central America: Haiti, Réunion, Martinique, Antigua and Barbuda, Puerto Rico, and El Salvador;

in Southwest Asia: the Islamic Republic of Iran, the Syrian Arab Republic, and Iraq;

in Africa: Algeria, Mauritania, the Niger, Somalia, Kenya, Burundi, and Lesotho.

• However, *19 countries (see Figure 8) would still remain critical even with high inputs* and must look to secure sources of foreign exchange and slower population growth — as well as agricultural development — to balance food needs and food provision.

Raising inputs is not just a matter of "modernizing" agriculture. *Methods appropriate to national social and ecological conditions must be used*. Fertilizer use must be increased — but complemented by organic recycling and biological fixation of nitrogen. Achieving higher input levels will also involve institutional changes, improved equity in agrarian relations, land distribution, availability of inputs to small farmers, and participation.

Changes in the pattern of crops so that they are more suited to local soil and climate conditions can produce considerable gains by themselves: a change from the present crop mix to the optimum in Africa could increase the low-input population-supporting capacity in year 2000 by *58 percent*. *Conservation measures* alone — with no increase in other types of input — could reduce by seven the number of critical countries in Africa with low inputs.

Slowing the momentum of population growth

In reducing the potential food requirements, every year of action or inaction counts. Small differences in population growth rates now build up to enormous long-term differences in the size of populations. For Africa, the UN low projection for 2025, for example, is 1 109 million — 430 million less than the *medium* projection. The difference is equal to an entire 1979 Africa.

Nine countries would be removed from the critical lists if they could reduce their population growth rates in line with the UN's low projection.

Six would no longer be critical for low inputs in the year 2000: Benin, Mali, Sierra Leone, Swaziland, Zimbabwe and the Dominican Republic.

Three would cease to be critical for intermediate inputs: Mauritania, the Syrian Arab Republic and Bangladesh.

None of the countries critical for high inputs in 2000 would cease to be critical.

However, all critical countries could benefit from reduced population growth rates: in all cases, the increases in agricultural effort required to meet food needs would be less strenuous, or the additional imports required would be lower. By the year 2025, for example, on the *medium* population projection, Africa would need to raise its food production by 144 percent above the low-input level to meet its food requirements, or to import 59 percent of its food needs. On the low projection, it would need to raise production by only 72 percent, or to import 42 percent of its requirements.

Measures that have proved effective in slowing population growth include wide-reaching family-planning programmes, combined with emphasis on the development of human resources through improved health care and education, improvements in the status of women, and redistribution of income in favour of poorer social groups.

International dimensions

North-South. Achieving higher input levels — or financing imports to meet the food deficits that will still remain — will involve access to technology, and to stable sources of foreign exchange. Securing the latter will involve reliable trade of manufactures or agricultural commodities, new arrangements for primary commodities to ensure more stable and equitable prices, and access to adequate credit.

Increased aid may also be required for those critical countries among the high-risk categories (see Fig. 8) whose own technical and financial resources are inadequate to stimulate the efforts required in balancing their populations and possibilities of food production or purchase. In a number of cases it is clear that aid, including food aid, will be indispensable in bridging the period before land, food and populations can be balanced.

South-South. The study shows that there is potential for greatly increased self-reliance in food within developing countries as a whole, and within most regions.

In many cases, potentially critical countries are adjacent to countries that could produce considerable surpluses. Utilizing this potential would stimulate intra-regional trade, and increase food security by reducing dependence on food production and stocks from distant countries.

The next 40 years will constitute a turning point in human history: the period when it will be decided whether populations can be accommodated to national resources without widespread suffering, disruption or degradation of the environment. The outcome concerns not only the potentially critical countries but the whole human race, and it depends on our present actions or omissions.

1. Origin and methods

The Land Resources for Future Populations Project was based on no less than 20 years of preparatory work by the Food and Agriculture Organization of the United Nations and cooperating countries and agencies.

Two major projects provided the foundation work. The first of these was the *FAO/Unesco Soil Map of the World*, on which work started in 1961. The regional maps, published between 1971 and 1981, provided the first detailed appraisal of the world's soil resources in a common, internationally accepted language (Ref. 1).

A second study, the *Agro-ecological Zones Project*, used the information of the Soil Map, combined with climatic data, to assess the extent of lands in developing countries that were suited to the production of various major crops (Ref. 2).

Publication of the first results of this study, for Africa, attracted the attention of the United Nations Fund for Population Activities. UNFPA wanted to know if it was possible to calculate the potential populations that the land resources of developing countries could support, and to identify critical areas and countries that would be unable to support their present or projected populations from their own lands.

As the answer to both questions was "yes", the UNFPA commissioned FAO to undertake the study, in collaboration with the International Institute for Applied Systems Analysis.

The methodology of the analysis was extremely complex — a full description can be found in the technical report *Potential Population-Supporting Capacities of Lands in the Developing World* (Ref. 3). What follows is a non-technical summary.

The Soil Map provided the essential data on soils, texture, slope and other physical characteristics that affect management and productivity, such as the presence of hard rocks close to the surface.

A separate inventory of data on climate had been produced for the Agro-ecological Zones Project. This covered rainfall, temperature and radiation data and divided the developing world into major climates. It also introduced the concept of a length of growing period zone. This is an area within which rainfall and temperature conditions are suitable for crop growth for a given number of days in the year — for example, 1-74 days, 75-89 days, 90-119 days, and so on. Map 2 illustrates a section of the climatic inventory.

The crucial step was to superimpose the climatic map on the soil map (See Map 2). This produced a fine mosaic of unique land units, within which soil and climatic conditions were known. The conditions in each cell could then be compared with the known requirements of different food crops to assess which crops were most suitable and what the potential productivity might be. The total number of cells in the mosaic was very large — for Kenya alone, for example, they numbered 1 213. A total of 117 countries were studied, covering almost all developing countries except China, Mongolia, the Democratic People's Republic of Korea, the Republic of Korea and some very small territories where climatic data were insufficient.

The assessment considered the potential for the rain-fed production of 15 major food crops: millet, sorghum, maize, rice, wheat, barley, soybean, *Phaseolus* bean, sweet potato, white potato, cassava, groundnuts, banana/plantain, sugar cane, oil palm — as well as for grassland supporting livestock in areas not suited for food crops.

Because inputs such as fertilizers have such a strong effect on yields, the analysis was made for three levels of inputs:

a *low level*, roughly corresponding to a rural area in a least developed country, using no fertilizer, pesticide or herbicide, with traditional crop varieties and existing cropping patterns, and no long-term conservation measures;

an *intermediate level*, using a basic package of fertilizers and biocides with some improved crop varieties, simple long-term conservation measures, and existing crop mixes on half the land and the most productive crop mix (in terms of calorie-protein yield) on the other half;

a *high level*, involving full use of fertilizer and biocides, improved crop varieties, conservation measures and the best mix of crops on all the land.

Deductions from the land area were made for non-agricultural uses such as settlements, industry and transport, as well as of land required for irrigation and for rest periods (fallow periods). Under the low and intermediate input levels, land loss and reduced productivity due to soil erosion were also taken into account. Deductions from the potential production were made to allow for seed, and for post-harvest losses assumed to be a uniform 10 percent.

The present and projected production from irrigated areas was taken from FAO's *Agriculture: Toward 2000* study. This was then added to the rain-fed production to calculate the potential calorie and protein production for each length of growing period zone, in each climate, in each country, for the three levels of input. For the intermediate and high input levels, a minimum protein availability was required from each zone.

The next task was to calculate how many people this potential production could support. This was done by dividing the total potential calorie production by the per caput calorie intakes recommended by FAO and WHO for each country.

These potentials were compared with existing and projected populations, to identify which areas and countries seemed likely to have insufficient land resources to meet the food needs of their present or future populations.

The calculations were made for two time frames: for the year 1975, subnational population data were available, and this allowed the identification of critical zones *within* countries, as well as critical countries and regions; for the year 2000, due to the difficulty of predicting population movements within countries, reliable projections exist at national level only, hence only the critical countries and regions could be identified.

2. The context of agriculture in developing countries

The major findings on population potentials are reported in the next chapter. The preliminary research, however, produced a great deal of background data which help to outline the framework for food production in developing countries.

Climate

Climate is a crucial factor affecting potential productivity, even in determining which crops can and cannot be grown.

Temperature is one component: crops require specific temperature conditions, and cannot grow well, or sometimes cannot grow at all, if these condition are not met. Provided temperature conditions are satisfactory, the degree of success in the growth of a crop under constant management depends largely on how well its optimum length of growth cycle fits in with the period when sufficient water is available for growth.

Other climatic constraints, known as agro-climatic factors, also affect yields: they include factors such as variability of rainfall, workability of soil, ease of drying crops, and pests and diseases which, like plants, flourish under specific climatic conditions.

Climatic factors have a tremendous impact on crop potentials. Cassava, for example, attains its optimum yields as a food crop when the growing period is between 270 and 330 days. For shorter periods, rainfall is inadequate. For longer periods, diseases such as leaf mosaic blight, and pests such as white flies and nematodes depress productivity. In very wet areas, productivity can

FIGURE 1

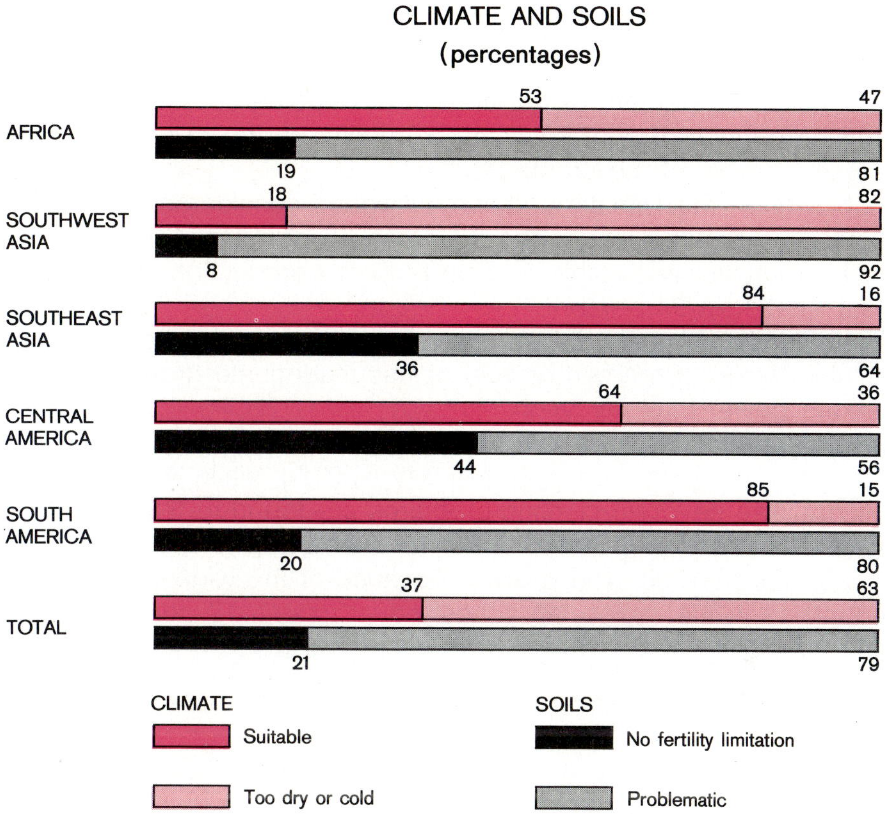

be only 38 percent of maximum. In dry areas it can fall below 10 percent.

By contrast, lowland maize does best in zones with growing periods between 150 and 240 days. In drier zones, problems such as rainfall variability and drying of silks can cut productivity to as low as 10 percent of the maximum; in wetter areas, borers, leaf spot, leaf blight and streak virus, wet cobs and unworkable soils can also pull productivity down to as low as 10 percent of the maximum.

Rain-fed wheat gives best results in the cool subtropics, and is unsuited to the warm tropics. Rain-fed rice, by contrast, performs best in all but the very wettest of well-watered areas, and does not do at all well when the growing period is below 150 days (Refs 4 and 5).

Overall, 63 percent of the area of developing countries enjoys growing periods of over 75 days, conducive to the rain-fed growth of one or more of the crops studied.

However, the climatic endowment varies considerably between regions. In South America, no less than 85 percent of the land area is climatically suitable for rain-fed crop production, and in Southeast Asia 84 percent.

In Central America, 64 percent of the area is climatically suitable, while 36 percent is arid or has inadequate growing periods (fewer than 75 days).

Climatic constraints are more widespread in Africa. Only 53 percent of the land area is climatically suitable, while 47 percent is too dry. The severest climatic problems are found in Southwest Asia where only 18 percent of the land area enjoys a climate suitable for rain-fed crop growth. Because of the widespread high mountains in the region, an almost equal extent, 17 percent, is too cold for crop growth, while the remaining 65 percent is too dry.

Soils

Soil conditions also have a great influence on potential crop production. Indeed, in many cases, serious soil limitations such as shallow depth, poor drainage or salinity are so severe that they can turn areas with excellent climatic potential into non-productive land, or at best poor pasture land.

Overall, only 21 percent of the area of the five regions enjoys soils with no inherent fertility limitations. They vary considerably in their soil resource endowment. Once again, Southwest Asia and Africa are the least well provided. In Southwest Asia, less than 8 percent of the land area enjoys soils with no inherent limitation to fertility. Just over a third of the area suffers from semi-desert and desert soils, with shallow soils covering another 27 percent of the area.

In Africa, only 19 percent of the land area has no inherent soil fertility limitations. Coarse-textured soils and desert and semi-desert soils cover 36 percent of the area.

One fifth of South America's land has no serious soil fertility limitations. The major constraint on the remaining area is severe fertility limitation affecting 41 percent.

Southeast Asia and Central America have the most favourable soil endowment. Thirty-six percent of Southeast Asia's land area has no inherent fertility limitations. In this region, the major constraint is severe fertility limitation, affecting a quarter of the land area. No less than 44 percent of the area of Central America has no inherent fertility limitation. The major constraint here, affecting 22 percent of the area, is shallow soil. (See Fig. 1.)

The land base

Those areas where both soil and climate are suitable for crop growth constitute the potential land base for rain-fed crop production. Not all of this potential, however, will actually be available. Some will be used for *irrigated crop production*. For this, the land resources project took the data of the previous FAO study *Agriculture: Toward 2000*, according to which the irrigated area in developing countries could expand from 95 million hectares in 1975 to 148 million hectares in 2000 (Ref. 6).

This is a realistic growth rate, somewhat slower than in the recent past. By the end of the century, the area equipped for irrigation would constitute 34 percent of the total arable area in the Far East and 27 percent in the Near East, but only 7 percent in Latin America and a mere 2 percent in Africa.

Irrigation is likely to account for a high proportion of the potential population-supporting capacities in three of the five regions. By 2000, at the low-input level, irrigation will account for 64 percent of the capacity in Central America, 73 percent in Southeast Asia and no less than 89 percent in Southwest Asia. In

Africa, however, it would account for only 20 percent of the potential production and in South America only 13 percent. The contribution at higher levels of inputs is proportionally lower, but this is partly because the study allowed only for the projected increase in inputs to irrigated land, as assessed by *Agriculture: Toward 2000,* and did not assume that irrigated inputs would be any higher at the higher input levels.

As the population rises, a growing area will be taken up by *non-agricultural uses*, such as housing, transport, industry, mining, conservancy and recreation. The proportion of land taken up by such uses varies widely, but after comparing cross-national surveys, the present study decided to allow a standard area of 0.05 hectare per person. On this basis, in 1975, a total of 98 million hectares, 1.5 percent of the total land area, would be taken up for non-agricultural purposes in the five regions. By 2000, the figure would rise to 180 million hectares, or 2.8 percent of the total area. The share would be relatively low in Africa, Southwest Asia and South America, but in Central America it would amount to 4 percent of the land area in 2000, and in Southeast Asia to 11 percent.

A much more serious threat to the potential land base for food production is *land degradation.* This can result from processes such as soil erosion or soil loss by wind or water, salinization or alkalinization, depletion of plant nutrients and organic matter, deterioration of soil structure, and pollution.

The land resources study developed a methodology for assessing the potential impact of land degradation due to wind and water erosion. It used indices of local susceptibility to wind and water erosion, based on the data on soil, slope, texture and estimated vegetation cover, to estimate the probable rates of soil loss that might occur over the long term if no conservation measures at all were undertaken. It was assumed that the full rate of this loss would occur at the low-input level. At the intermediate level of inputs, which involves some basic conservation measures, the rate of loss would be halved, while at high-input level, where full preventive measures would be taken, losses would be within acceptable limits, without depressing production.

The results of the assessment gave cause for grave concern

FIGURE 2

THE DEGRADATION THREAT: LOSSES OF RAINFED CROPLAND
(million ha)

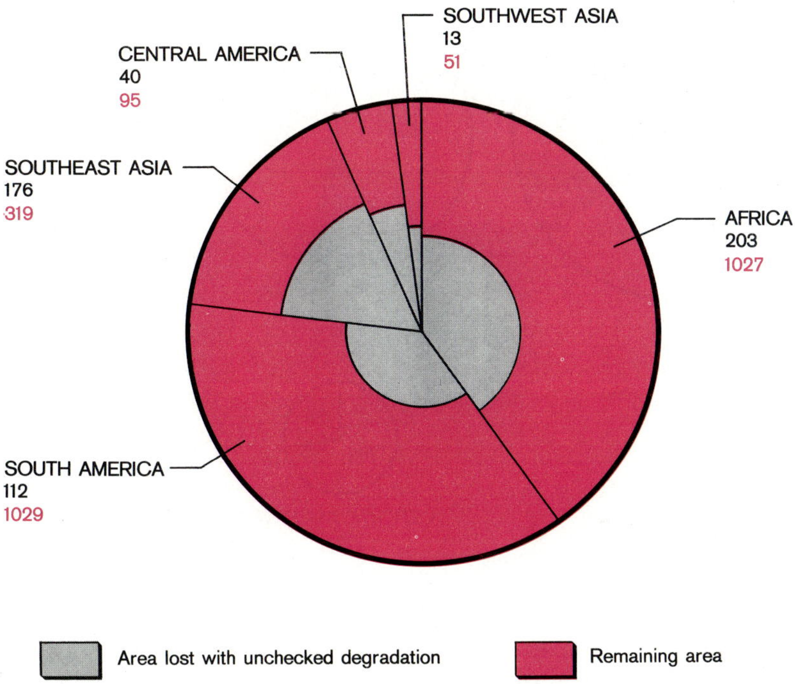

over the consequences of unchecked degradation (see Fig. 2: degradation threat). Without conservation measures, the area of rain-fed cropland over the five regions as a whole would shrink over the long term by 18 percent, or 544 million hectares — an area greater than the entire potential cropland of Southeast Asia.

In South America, the loss would be only 10 percent, but it would amount to 16.5 percent in Africa and to 20 percent in Southwest Asia. In Central America, 30 percent of potential cropland would be lost, and in Southeast Asia as much as 36 percent.

Much of the cropland that remained would also lose fertility

FIGURE 3

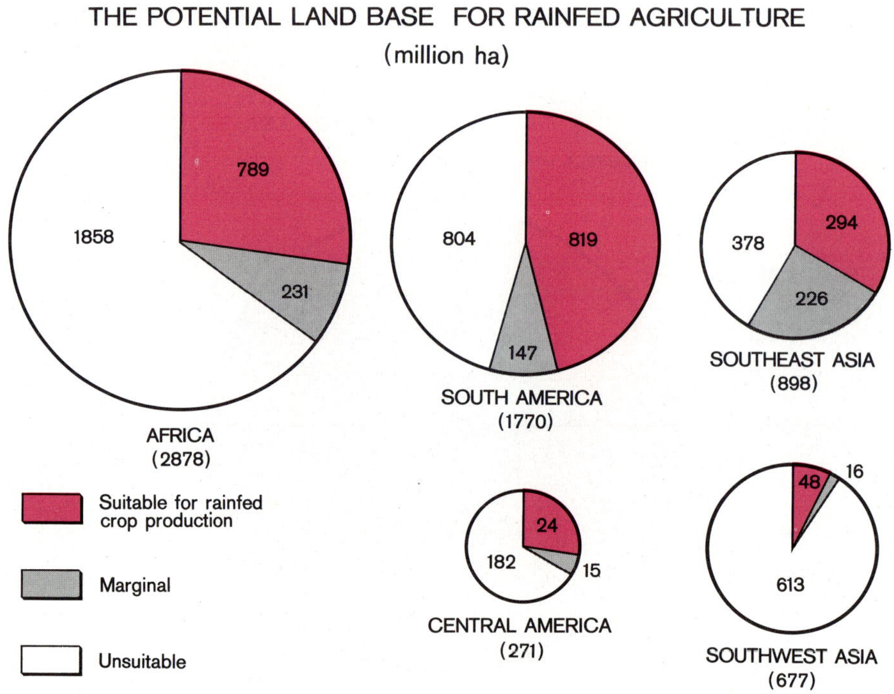

owing to loss of topsoil, so the total loss in productivity of rain-fed cropland would amount to a daunting 29 percent. However, much of the land lost to crops would still be usable as grassland supporting livestock. Thus, the loss of overall land productivity would be lower, but still very serious, averaging 18.5 percent for all five regions, but ranging up to a quarter in Africa and Central America.

The overall potential land base for rain-fed crop production in developing countries is very large (see Fig. 3). Just how large it is depends on whether it is restricted to suitable land, or, as in this study, includes land that is poorly suited but still capable of giving some yield. The potential also varies according to the level of input: more land is lost to degradation with low inputs, but for the higher input levels certain types of land, such as steeply sloping land, are assumed to be unsuitable for cultivation.

3. The findings

The detailed character of the land inventory, described in Chapter 1, allows a wide range of results to be calculated, from the small area of the agro-ecological cell, through climates and length of growing period zones, to countries and regions. The results will be presented in this chapter.

Before considering them, however, it is important to bear in mind the limitations of the assessment. The aim of the study is to provide a broad evaluation of the potential for food production of the entire land resources of developing countries. Hence, *the results imply the use of all potentially cultivable land*. In Southwest Asia and Southeast Asia, and other countries where land reserves are becoming scarce, this is not very different from the present cropland, but in the regions of Central America, Africa and South America the total potentially cultivable land is between two and seven times greater than the present cultivated area.

The study assumes that the *whole of the potentially cultivable land is used to grow nothing but food crops, or grassland supporting livestock* if it is too marginal for crops. No allowance is made for other essentials such as fibres, or vegetables and fruit, or for the use of cereals as livestock feed, or for non-food cash crops. Forestry for fuelwood or timber is restricted to lands unsuitable for crops or grazing.

Only the average recommended calorie intakes identified by FAO and WHO are taken into consideration. These relate to the present age structure of the countries concerned, which include a high proportion of children with lower calorie needs. As population growth rates slow down in each country, this proportion will decline and the national average calorie requirement will rise. In addition, as incomes rise, calorie intakes above the minimum are usually preferred. *No allowance has been made for inequality in the distribution of food*. A country which may be able to provide from its own lands the recommended calorie intake as an *average* for all of its population may still have substantial numbers suffering hunger. All of these factors would tend to reduce the potential carrying capacity below the results quoted in this chapter.

Several other factors might tend to improve them. First, *potential contribution of marine and inland fisheries and aquaculture* to calorie and especially protein supply is not accounted for. Inclusion of fisheries would make a significant difference to the results for several countries and regions. Second, *the contribution of irrigation* could be higher than that allowed for here, if irrigation systems, and the inputs used in them, were to increase faster than the rate projected by *Agriculture: Toward 2000*. Conversely, the contribution would be lower if these rates were not achieved. Third, the land was assessed *in its present condition*: future major improvements affecting suitability, such as levelling, terracing, food control, drainage, would increase the potential.

The study could not take into account certain imponderable and unpredictable factors which could have a negative or a positive effect on production. *Major technological breakthroughs*, for example, in the use of biological nitrogen fixation by cereals, might increase the potential yields achievable at given input levels. Development and diffusion of crop varieties, specially adapted to adverse conditions such as salinity, low rainfall, etc., would also increase the potential output. *Climatic factors* such as changes in global temperatures and carbon dioxide levels, or the location of major rain-belts, could have positive or negative effects, varying from one region to another.

Finally, the results, and indeed the whole study, do not propose or imply that self-sufficiency in food is a necessary goal for every country. The study merely assesses the *potential* for food production within each country, from its own lands, in the hope that this information will prove useful to governments and

FIGURE 4

POPULATIONS OF CRITICAL COUNTRIES
(millions)

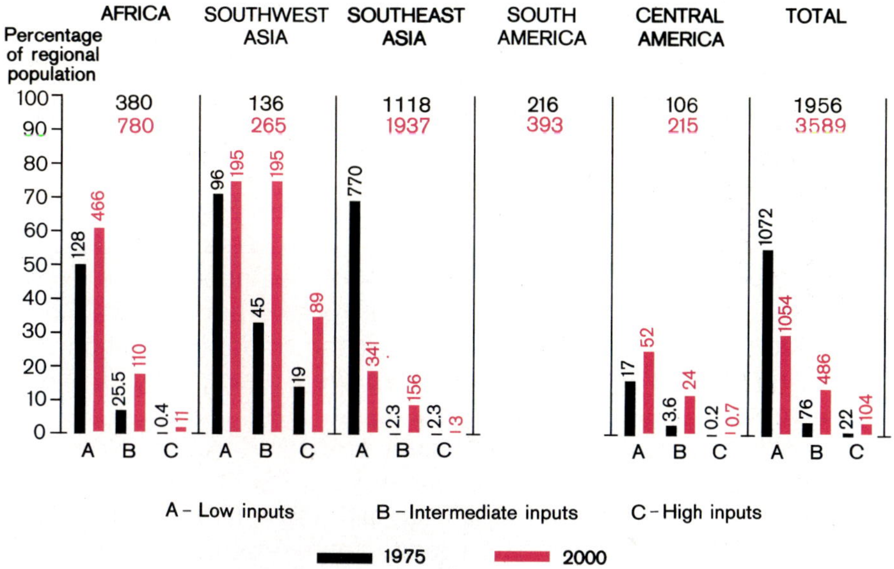

NUMBER OF CRITICAL COUNTRIES

		Africa	SW Africa	SE Asia	S. America	C. America	TOTAL
Total countries		51	16	16	13	21	117
1975	Low	22	15	6	—	11	54
	Int.	7	12	1	—	4	24
	High	2	9	1	—	1	13
2000	Low	29	15	6	—	14	64
	Int.	12	15	2	—	7	36
	High	4	12	1	—	2	19

planners, whether they wish to pursue self-sufficiency as a goal or not. The study does not take into account the role of trade, international loans or aid, which may, or may not, enable various countries which cannot produce sufficient food from their own lands to buy whatever imports they need. Clearly, there are some countries, with established or rapidly growing manufacturing industries, with long-term reserves of fossil fuels, or profitable and reliable exports of cash crops or minerals, who will have few problems in this respect. Equally clearly, there are others whose ability to import their potential food deficits is insecure, and for whom the results of this study may provide the opportunity to plan a sustained balance between populations, land resources, and sources of foreign exchange.

As explained in Chapter 1, the results are given for three levels of input: low, intermediate and high. The original study did not attempt to place individual countries or regions on the input scale, but it may help readers to interpret the following results if some indication is given.

One way of assessing the input level is by fertilizer use. The low level of inputs implies the use of no fertilizers whatsoever; the intermediate, around 80 kg of nutrient per hectare per crop for cereals; and the high level, around 160 kg per hectare. In 1980, none of the regions had attained the intermediate figure. Developing market economies in Africa were using only 9.7 kg of nutrients per hectare — and, except in North Africa, a very large proportion of that goes on cash crops for export. Near Eastern countries were applying an average of 34 kg per hectare and Far Eastern ones 38 kg, although in both cases a high proportion went to irrigated land. Latin American countries applied an average of 46 kg, but in most of them a high proportion of this went on cash crops. Thus, to give a less distorted picture of fertilizer use with food crops, the African and Latin American figures should be reduced by at least half. Figure 4 (which includes this reduction) shows the levels each region had reached, taking the difference between low inputs and intermediate inputs as 100 percent (Refs 7 and 8).

Another indicator of the level of agriculture is the average yields achieved. The study's results are based on the known growth characteristics of crops under given conditions of climate, soil and input: the crop yields assumed can then be compared with the actual yields achieved (see Appendix 3; see also Ref. 10).

On average, African yields for most crops in 1979-81 fell below what could be achieved, even with low inputs, in favourable environments. Hence, it seemed reasonable to place

FIGURE 5

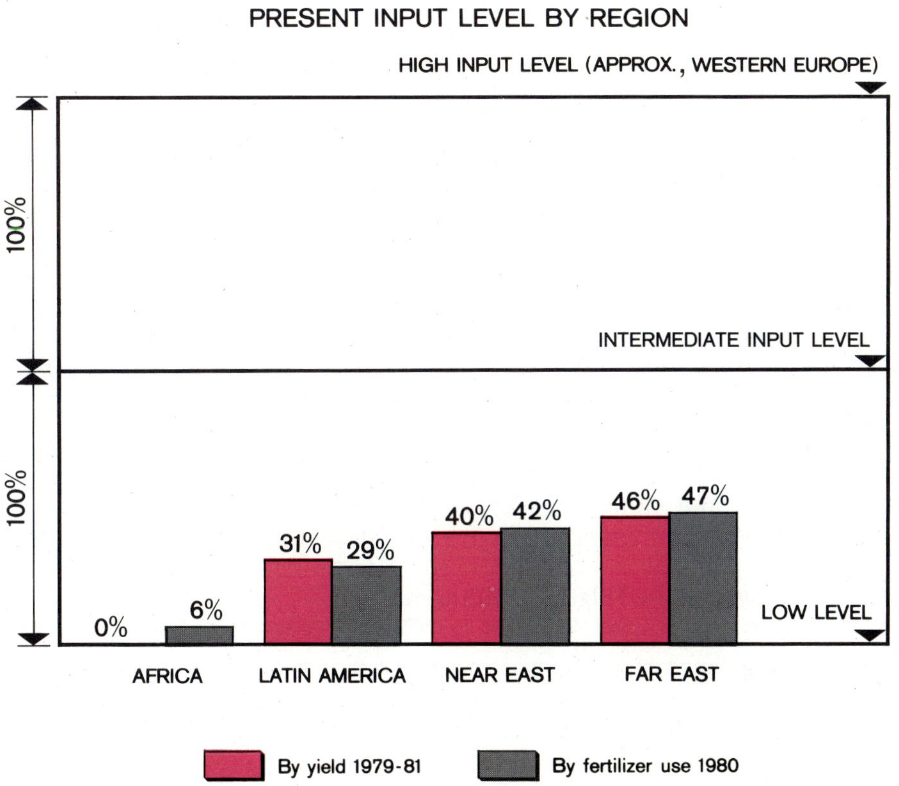

Africa at the low level of inputs. Comparing the yields of the other regions suggested that they had progressed some way toward the intermediate-input level, but none of them had reached it. Latin America was about one third of the way between low and intermediate, the Near East about two fifths of the way, the Far East just under half of the way (see Fig. 5). At the same time, many poor and marginal farmers and poor countries might be little above the low level. For comparison, using the same method, Western Europe was slightly above the high level (Ref. 12).

Results: the global overview

The most general level of analysis in the study looks at developing regions as a whole. This makes the very unrealistic assumption that surplus food and labour flow freely across national and regional borders, but it does give some idea of the potential for reducing developing countries' dependence on food imports from developed countries.

Taken together, the entire land resources of the 117 countries studied would have been able to support twice their total 1975 population of 1 956 million, using low inputs. With intermediate inputs, they could support almost seven times, and with high inputs, more than 16 times their 1975 population (see Table 1). By 2000, the total population of the five regions will have grown to 3 589 million. The potential supporting capacity will also have increased because of projected increases in irrigation, but not quite as fast as population.

At the low level of inputs, the situation for 2000 is extremely

Table 1. REGIONAL POTENTIAL/NEEDS RATIOS
(Potential population-supporting capacities divided by 1975 or 2000 populations)

INPUT LEVEL	AFRICA	SW ASIA	SOUTH AMERICA	CENTRAL AMERICA	SE ASIA	AVERAGE
YEAR 1975 RATIOS						
Low	3.0	0.8	5.9	1.6	1.1	2.0
Intermediate	11.6	1.3	23.9	4.2	3.0	6.9
High	33.9	2.0	57.2	11.5	5.1	16.6
YEAR 2000 RATIOS						
Low	1.6	0.7	3.5	1.4	1.3	1.6
Intermediate	5.8	0.9	13.3	2.6	2.3	4.2
High	16.5	1.2	31.5	6.0	3.3	9.3

tight. The entire potentially cultivable lands of the developing countries, more than three times the present cultivated area, could support only 1.6 times the expected population, leaving little margin for non-food crops or for population growth after the year 2000.

With intermediate inputs, four times the year 2000 population could be supported, and with high inputs, nine times. Both these figures would be adequate to allow for the expected future growth in populations, even up to the year 2100. Thus, in theory, with the application of higher input levels and the use of all land resources, the developing countries as a whole would be well capable of self-reliance in food supplies.

This global result, however, says nothing about the prospects for individual countries. The country level is probably the most significant level of analysis for the practical world, for, in the last analysis, each country has to feed its own population, whether by production from its own lands, or by paying for food imports with the proceeds of trade, loans or aid. The results for this level assume that food flows freely between length of growing period zones within each country, so that surplus zones can help to feed people in deficit zones.

The country results give considerable cause for concern. In 1975, there were 54 countries, out of the 117 studied, which would have been unable to feed their existing population from their own lands, using low inputs (see Fig. 4). These countries covered 2 221 million hectares — just over a third of the total land area — but they contained no less than 55 percent of the total population or 1 072 million people. The entire land resources of these critical countries were capable of feeding only 794 million people with low inputs — hence, 278 million, or more than one quarter, were in excess of the capacity. The situation improved considerably if intermediate inputs were used. Then, only 24 countries, with a mere 4 percent of the total population, would be critical. At the high-input level, only 13 countries, accounting for only 1 percent of the total population, would be critical.

By year 2000, the total population of the 117 countries is expected, on the medium UN projection, to increase by 83 percent, to 3 589 million. For most countries, this will mean a worsening of the land resources and population balance. The number of critical countries with low inputs will rise to 64. Twenty-eight of these would cease to be critical if they had reached the intermediate-input level, while another 17 would no longer be critical if they applied high inputs; but 19 countries would remain critical even with high inputs.

Because India will cease to be critical at the low-input level, owing to planned increases in irrigation production which will increase its potential, the total population and extent of critical countries for low inputs would actually decline slightly by 2000, but the excess of the population over the supporting capacity would rise to 503 million, no less than 48 percent of the population of the critical countries. Moreover, the total population of countries which will still be critical with intermediate inputs will rise from 76 million in 1975 to 486 million.

Threatened zones and climates

The results are also analysed in the study according to length-of-growing-period zones within each country. Because of problems in forecasting population movements *within* countries, this level of assessment was made only for 1975. It is based on the somewhat unrealistic assumption that there would be no movement of surplus food between zones in the same country — but it is useful in pinpointing the problematic areas within countries where land resources cannot meet the needs of local populations. These are often areas where the land is overexploited and the threat of land degradation is greater. The length-of-growing-period zone results, therefore, give a good idea of the real extent, on the ground, of the imbalance between land resources and populations.

In 1975, a total of no less than 2 450 million hectares, 38 percent of the total land area, was carrying more people than could be fed with low inputs; 1 165 million people, or 60 percent of the total population of the 117 countries, lived in such zones, which were capable of feeding less than half that number, or only 563 million people.

These critical *zones* are more problematic than the critical *countries*, in that they are less responsive to the application of higher input levels. With intermediate inputs, 26 percent of the total area, containing 20 percent of the population, would still be critical. Even with high inputs, the critical zones would still occupy 22 percent of the area and contain 11 percent of the population.

The critical zones, delineated in Map 1, are clearly areas that must command particular attention in the planning of agricultural development and population distribution policies and the allocation of aid.

Why are there critical zones?

Analysis of the results by climate helps to suggest the principal reason for the existence of these critical zones. Apart from irrigated areas, the highest potential population-supporting capacities are found in the warm tropics and subtropics, and in the more humid zones with length of growing periods above 240 days.

However, the distribution of human populations suggests that many people may have a certain reluctance to settle in hot and humid areas. The reasons for this were not explored, but major factors may be the higher incidence of human and animal diseases, the difficulties of clearing and cultivating tropical forest areas with traditional crops and technologies, and the inaccessibility and lack of infrastructure in such areas. On average, the highest *actual* population densities are found in the moderately cool tropics, the warm subtropics, and in the zones with medium-length growing periods (between 90 and 240 days).

The greatest excess of present population over potential supporting capacities is found in the cool and cold tropics and subtropics, and in the zones with growing periods between 75 and 150 days (see Fig. 6).

These results indicate a degree of imbalance between the distribution of human populations and the potential of the land

FIGURE 6

THE CLIMATIC IMBALANCE
(density persons per ha)

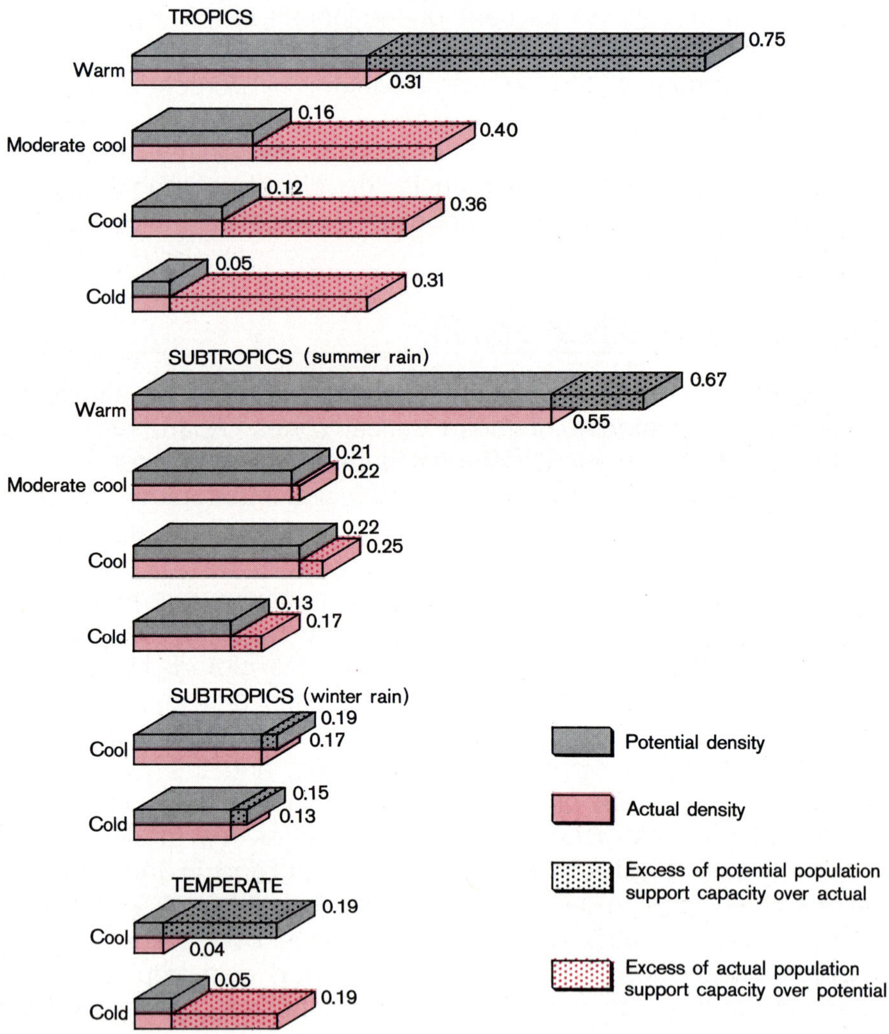

resources. In countries which straddle lower and higher potential zones, such as most Latin American countries, government measures may be able to reduce this imbalance. In some areas such as East Africa, however, national boundaries confine some countries to mainly low potential zones, while neighbouring central African countries enjoy mainly high potential zones.

Regional survey

The country results were also aggregated on a regional basis. This level of analysis assumes a free flow of surplus food production and labour across national boundaries within a region. Although this, again, is not a realistic assumption, it does provide a clear indication of the potential for regional self-reliance in food production and, therefore, of the possibilities of intra-regional trade in food.

These aggregate regional results are given below, together with the individual country and length-of-growing-period zone results by region (see Table 1 and Map 1).

Africa

Africa's prospects of feeding its future populations seem most clouded. The continent has a massive land area of 2 878 million hectares (excluding South Africa), but 47 percent of the area is too dry for rain-fed agriculture and only 19 percent of soils have no inherent fertility limitations. Despite this, Africa has a very large area of potential rain-fed cropland — 789 million hectares, not including marginal land — most of this in the humid tropics. In 1975, only 168 million hectares were cultivated.

Africa's yields in 1979-81 corresponded roughly to the low level of inputs, while fertilizer use averaged only 10 kg per hectare, far below any other region.

The existence of large unused land reserves meant that the region *as a whole* was capable of supporting 1 120 million people in 1975 with low inputs — three times the actual population of 380 million.

With intermediate inputs, the land resources could support almost 12 times the 1975 population, and with high inputs almost 34 times. However, land reserves are very unevenly distributed, with central African countries very favourably placed, while many countries in North Africa, the Sahel and East Africa are already using a very high proportion of their potentially cultivable land.

This unevenness shows up clearly in the zone and country results. Almost half the total land area, or 1 355 million hectares, could not support its 1975 populations with low inputs; 184 million people — 48 percent of the regional population — lived in such zones, which were actually capable of feeding only 79 million people.

The critical zones extend across the Maghreb, in mountain and coastal areas. They stretch across the Sahel between latitudes 20°N and 12°N, taking in the northern savanna areas of Ghana, Togo and Nigeria. The most densely populated areas of Kenya are critical, along with a densely settled crescent along the great lakes from southeast Uganda through Rwanda and Burundi to parts of Malawi. In the south, critical zones straddle the continent, from the coast of Angola, through Botswana and Lesotho, to southern Mozambique.

Raising inputs to the high level reduced the critical zones to 30 percent of the area, containing only 12 percent of the population. At this level, the only large critical areas are found in the Atlas mountains, Mauritania, Somalia, southern Ethiopia, Kenya and Rwanda.

The country results are not encouraging. With low inputs, 22 of the 51 countries in the region would have been critical in 1975, with a population of 188 million (half of the African total). Of these, 67 million were in excess of the critical countries' supporting capacity. Use of intermediate inputs brought down the number of critical countries to seven, while high inputs left only two critical.

By the year 2000, the situation becomes very much more serious. The population is expected to more than double, to 780 million — averaging a growth rate of 2.9 percent a year between 1975 and 2000, the fastest of any region — while

FIGURE 7

AREA AND POPULATIONS OF CRITICAL ZONES (Year 1975)

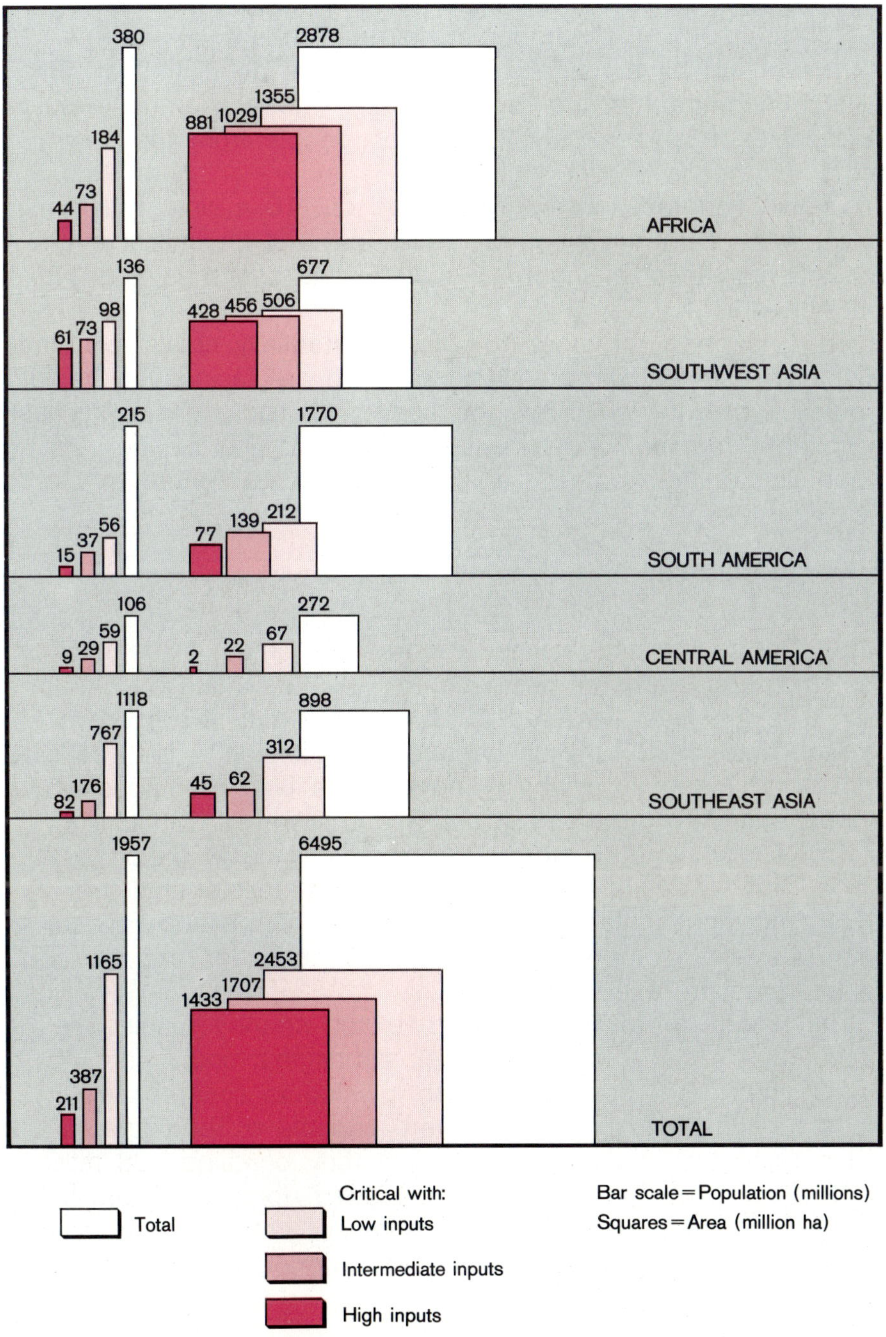

irrigation potential is not expected to expand very much.

By year 2000, the number of countries critical at the low input level will have risen to 29, with a total population of 466 million (60 percent of the regional total). These countries would be capable of feeding only 210 million people from their own lands.

Twelve countries, with a population of 110 million, will remain critical with intermediate inputs, while four will be critical even with the high level of inputs. (For individual countries, see Fig. 8 and Appendix 1.)

The prospects for most of Africa's potentially critical countries must give rise for concern. Most of them are facing very rapid population growth, while few are seeing industry develop as fast as in other regions. The capacity of many to finance the importation of their potential food deficits is not secure.

Southwest Asia

This is the region where land resources are most inadequate to meet the food needs of populations. Both in climate and in soils, it is the least favoured of all regions: only 7 percent of the land area of 677 million hectares is suitable for rain-fed crop production. Taken as a whole, the region's land resources could support only 107 million of its 1975 population of 136 million, using low inputs. With intermediate inputs, it could support 172 million, and with high inputs, 268 million, almost double the 1975 population. In fact, around 1980, the region had probably reached a level somewhat under half way between the low and the intermediate levels.

The length-of-growing-period zone and country results are unfavourable. At the low-input level, zones that could not support their 1975 populations with low inputs covered three quarters of the land area. They were capable of feeding only 33 million people at this level, but in fact they contained 98 million — 72 percent of the regional total. Indeed, for this region, it is simpler to mention the areas that were not critical — the only sizeable ones for low inputs being the interior and northeast of

FIGURE 8

THE CRITICAL COUNTRIES AND TERRITORIES, YEAR 2000

		LOWER RISK (Scoring 5 or less)	HIGHER RISK (Scoring 6-7)	HIGHEST RISK (Scoring 8-9)
COUNTRIES ABLE TO SUPPORT LESS THAN HALF YEAR 2000 POPULATIONS	HIGH INPUTS	NETHERLANDS ANTILLES BARBADOS CAPE VERDE		RWANDA DEMOCRATIC YEMEN YEMEN
		LEBANON ISRAEL MAURITIUS SINGAPORE	JORDAN	
		KUWAIT QATAR UNITED ARAB EMIRATES OMAN SAUDI ARABIA BAHRAIN		AFGHANISTAN
	INTERMEDIATE INPUTS	REUNION MARTINIQUE ANTIGUA	LESOTHO MAURITANIA	HAITI BURUNDI SOMALIA
			EL SALVADOR KENYA	
			IRAQ NIGER ALGERIA	
	LOW INPUTS	GUADELOUPE WINDWARD ISLANDS	NAMIBIA	COMOROS ETHIOPIA UGANDA
				NIGERIA
COUNTRIES ABLE TO SUPPORT 50-99% OF YEAR 2000 POPULATION	INTERMEDIATE INPUTS	PUERTO RICO		
			BANGLADESH	
		IRAN	SYRIA	
	LOW INPUTS	BAHAMAS VIET NAM	BOTSWANA BHUTAN SIERRA LEONE SWAZILAND BENIN	UPPER VOLTA MALAWI TOGO MALI
		DOMINICAN REP. JAMAICA ZIMBABWE SRI LANKA PHILIPPINES	MOROCCO SENEGAL GUATEMALA	
		TRINIDAD AND TOBAGO TUNISIA		

Indicates manufactured exports of over US $ 100 million per year, 1979 or 1980

Indicates significant wealth in fuel minerals

N.B. Western Sahara, critical with high inputs, is not included for lack of data

Turkey. The use of increased inputs improved the situation, but not dramatically — even high inputs left 63 percent of the land area critical, with 45 percent of the regional population.

With low inputs, 15 of the 16 countries in the region could not feed their 1975 populations from their own land resources: 96 million people lived in these countries, and 34 million of them were in excess of the supporting capacity. The only non-critical country in the region for low inputs was Turkey. Intermediate inputs reduced the number of critical countries to 12, but even high inputs still left nine countries critical.

By the year 2000, the situation will have grown even more serious. The main reason is that the population is expected to grow rapidly, from 136 million to 265 million, an increase of 94 percent, or 2.7 percent a year. The potential for food production will also grow, owing to projected increases in irrigation, but not so rapidly.

As a result, the region as a whole would be unable to feed its year 2000 population with low or with intermediate inputs, and even with high inputs could feed only 23 percent more than its expected population. Allowing one third of the land for non-food crops and food consumption factors (see next chapter), this means that the region will be critical by year 2000 even with high inputs, and large increases in population are expected thereafter.

In 2000, only one country, Turkey, would be able to feed its expected population from its own lands with low inputs. The other fifteen countries are critical for both low and intermediate input levels. With high inputs, the Islamic Republic of Iran, the Syrian Arab Republic and Iraq would cease to be critical but the other twelve countries, containing just over one third of the regional population, would be critical. Six of these countries are oil exporters. Another two — Israel and Lebanon — have low population fertility levels and diversified economies. Afghanistan, Democratic Yemen, Yemen and Jordan are less favourably placed: all four use low levels of fertilizer, have low average yields and very high population fertility. In all four, the year 2025 population is projected to be three or more times greater than in 1975.

Southeast Asia

This is the region where, because of the sheer size of its population, by far the largest absolute numbers of people affected by potentially critical land resources exist. Its climate and soil resources place it among the most favoured regions; one third of its total land area of 898 million hectares is potentially suitable for rain-fed crop growth. However, it is also pushing against the limits of those resources, cultivating 92 percent of them in 1975, although some countries in the southeast of the region still have substantial reserves.

With the low level of inputs, the region's land resources were capable of feeding 1 247 million people in 1975 — only 12 percent more than the actual total of 1 117 million. Allowing for non-food crops, this probably meant that the region as a whole was critical for low inputs.

However, Asia's agriculture is further advanced than in the other regions — as we have seen, the region is, on average, around half way toward the intermediate level of inputs. Using intermediate inputs, the region's lands would be capable of feeding just over three times the 1975 population, and with high inputs more than five times.

For 1975, only six of the region's 16 countries had insufficient land resources to feed their populations, using low inputs. Their total populations were 69 percent (770 million of the regional total). All six, however, were achieving yields closer to intermediate than to low level. Using intermediate or high inputs, only one country remained critical: the city state of Singapore, whose ability to finance its food imports with manufactured exports is well established.

The results for the length-of-growing-period zones for 1975 present a similar picture. With low inputs, zones covering 313 million hectares — just over a third of the total land area — would have been unable to support their 1975 populations of 767 million, 69 percent of the regional total. These zones covered most of India and Pakistan, the highland areas of Nepal and Bhutan, and the most populated areas of Sri Lanka, Viet Nam, Indonesia (Java) and the Philippines. Applying intermediate inputs, however, has a dramatic effect: the critical lands shrink to only 62 million hectares — 7 percent of the area — with a mere 16 percent of the population. They remain extensive only in northern Pakistan, highland Nepal, western Bangladesh and Java. High inputs reduce the area further, with a dramatic reduction even in Java.

Asia's population growth is slowest of the five regions. By the year 2000, the population of the countries studied will have risen by 73 percent, from 1 118 million to 1 937 million, averaging 2.2 percent a year. However, the large projected increases in irrigation will increase the potential supporting capacity faster than this, so that the region's lands would be able to support 27 percent more than their expected population, using low inputs. With high inputs, the region would be able to support three times its year 2000 population. With low inputs, six countries would be critical. Singapore would remain critical at all input levels. Only one other country, Bangladesh, would be critical for intermediate inputs, and even with high inputs would be able to support only 16 percent more than its expected year 2000 population, although this is projected to grow by another 50 percent by year 2025.

Central America

Just over a quarter of this region's land area of 272 million hectares is suitable for rain-fed agriculture. The substantial populations on the mainland highlands, and the fact that many countries here are small islands mean that land is not over-abundant. In 1975, about half the 74 million hectares that were potentially cultivable were actually being cultivated. Yields were equivalent to around a third of the way toward the intermediate level of inputs, and although fertilizer use is fairly high, a large proportion of that goes on export crops.

The region's population in 1975 was 106 million. With low inputs, the land resources were capable of feeding 1.6 times that number; with high inputs, no less than 11.5 times. This large potential, however, is unevenly distributed. Eleven of the 21 countries in the region were critical for low inputs in 1975, with a total population of 17 million, only 16 percent of the regional total.

Four countries — all islands — remain critical even with intermediate inputs, but with high inputs only one small island is still critical.

The length-of-growing-period zone results paint a more serious picture. One quarter of the land area, containing 55 percent of the total population, was in zones that were critical with low inputs. These zones were, for the most part, highland areas where subsistence agriculture, equivalent to low inputs, is prevalent. Use of intermediate inputs, however, greatly reduced the critical zones to just 8 percent of the area, with 27 percent of the population, while high inputs left only 0.8 percent of the area critical, although 9 percent of the population lived there.

The outlook for 2000 is less favourable. The population is expected to grow almost as rapidly as Africa's, more than doubling to 215 million, averaging 2.85 percent a year over the 1975-2000 period. The region as a whole could sustain 35 percent more than its expected population, with low inputs — which does not allow enough margin for non-food crops, etc. Intermediate inputs would permit it to feed 2.6 times the expected total, and high inputs, six times.

However, by the end of the century, the number of critical countries with low inputs will have risen to 14, with a quarter of the regional population. Seven of these, including El Salvador and Haiti, would remain critical with intermediate inputs. With high inputs only, two relatively prosperous countries, Barbados and the Netherlands Antilles, would remain critical.

South America

South America has the lowest incidence of critical areas and countries of any region, and the highest proportion of land area — 46 percent — that is suitable for rain-fed agriculture. The huge extent of the region — 1 770 million hectares — in relation to its population means it has abundant land reserves. In 1975, only 15 percent of the potentially cultivable area was being cultivated.

As with Central America, agricultural development is not so far advanced as in Asia. Average yields are closer to the low than to the intermediate level, while fertilizer use — although it appears relatively high at 46 kg per hectare in South America — often goes predominantly on export crops.

At the regional level, even with low inputs, the land resources would be able to feed almost six times the 1975 population of 216 million. With intermediate inputs, the potential supporting capacity rises to 24 times the 1975 population, and with high inputs, no less than 57 times. However, these results imply the use of *all* potentially cultivable land, and would involve the clearance for crops or pasture of more than half the 943 million hectares of forest and woodland, which is undesirable and unlikely, plus a major reduction in the use of potential cropland for grazing.

The country results for 1975 are favourable. Not one of the region's 13 countries is critical for any level of input, although Chile, with much smaller land reserves than the other countries, has a lower margin of potential surplus.

However, the continent is not free of problems. The length-of-growing-period zone results show that no less than 56.4 million people — more than a quarter of the regional total — live in zones unable to produce enough food to feed their 1975 populations, with low inputs. Apart from a few scattered areas along the dry Caribbean coast, and in inland Brazil, most of these critical areas are concentrated along the Andes, in cool and cold climates, where the bulk of the population does in fact farm around the low level of inputs. These are also zones with a high risk of desertification and degradation and an acute shortage of fuelwood.

Most of these Andean areas will remain critical even with intermediate inputs, when they will cover 8 percent of the land area with 17 percent of the regional population. They constitute a powerful human and ecological challenge. High inputs, however, could confine them to 4 percent of the area and 7 percent of the population.

By the year 2000, the regional population will have grown by 82 percent over the 1975 level, reaching 393 million, but the land resources will still be capable of feeding many times that population — between 3.5 times with low inputs and 31.5 times with high inputs. No countries will be critical at any level of input. The more scarcely populated areas have enormous potential: Guyana and Suriname could feed 39 times their year 2000 populations even with low inputs.

However, these favourable results do no more than indicate the ultimate potential of the land resource. In fact, the unequal pattern of landownership and land use in the continent — where the large holdings often cultivate less than a sixth of their area — keeps the performance well below the potential.

What do we know about the critical countries?

It is instructive to examine the characteristics of the countries that are potentially critical, to see to what extent they have the potential to overcome their problems — or to what extent the problem of inadequate land resources may be compounded by other development problems.

By the year 2000, 64 countries — more than half the total studied — would be unable to feed their expected populations from their own land resources with low inputs. While some of these countries — most clearly the oil exporters and Singapore — can hardly be said to have much of a problem in feeding their populations, many of the other countries suffer from an accumulation of agricultural and development problems which add up to a daunting challenge for their governments and for the international community.

Appendix 2 gives a detailed statistical analysis of these countries. Fourteen of them are landlocked. Another nineteen are islands. Twenty of them figure among the least developed countries. No less than three quarters of those for which data are available are expected to see their self-sufficiency rates in basic foods decline over the next two decades; 56 percent have average per caput daily calorie supplies lower than their recommended levels.[1]

[1] These figures are expressed as percentages for simplicity, as the total number of countries for which data are available varies according to the factor considered.

They are, for the most part, land-scarce countries — those with large land reserves tend not to show up as critical in the present study, which considers the potential of *all* usable land. Almost two thirds of the critical countries are cultivating more than 70 percent of their potentially cultivable area. Many of them are farming at low levels — 55 percent are applying less than 35 kg of fertilizer per hectare, while 69 percent are getting cereal yields below the average for developing market economies. An even greater proportion — 89 percent of those surveyed — face ecological problems, as indicated by an acute shortage or deficit of fuelwood on part or all of their territory.

Despite these present problems, most of the critical countries are likely to see fairly rapid population growth over the next 40 years. No less than 71 percent have high total population fertility rates, above 5,[2] while 60 percent have population fertility rates above 6. Barring cases of unusually rapid population fertility decline, such high rates usually mean a longish delay before population growth rates fall to modest proportions. Many of the countries may face problems in financing imports of food: 54 percent are relatively poor, with per caput annual incomes below US$1 000 in 1980.[3] Some 38 percent rely on agricultural products for more than 40 percent of their export earnings. Less than half have significant oil wealth or manufactured exports. Not all the oil-exporting countries can rely on long-term support from oil — in Trinidad, Bahrain, Qatar, Algeria, Syrian Arab Republic, the Islamic Republic of Iran, Oman and Nigeria, 1982 reserves would last only 40 years or less at 1982 production rates (Ref. 11).

Many of the critical countries, indeed, suffer from a dangerous combination of all the above problems. Each country can be scored on a scale of nine, giving one point for each of the following risk factors: land scarcity (using more than 70 percent of the potentially cultivable area); low fertilizer use

---

[2] The total fertility rate is the number of children that the average woman would have in a full reproductive lifetime if the pattern of that year continued. Europe's total fertility rate is around 1.9. — [3] The US$1 000 income threshold is chosen here as it appears to be a better indicator of the level below which the likelihood of a country having inadequate average calorie intakes becomes significant (see Appendix 2).

(below 35 kg per hectare); cereal yields below the developing market economy average of 1.6 tonnes per hectare in 1981; calorie supply below 100 percent of requirements; declining self-sufficiency in basic foods; fuelwood scarcity or deficit; population fertility rates above 5; per caput incomes below US$1 000; and agricultural products accounting for more than 40 percent of export earnings (see Appendix 2 for full details).

Twenty countries scored three or less, and do not face serious problems. Ten countries scored four or five — three of these had significant oil wealth. However, there were 34 countries, more than half the total, that scored six or more. Of these, no less than 23 were African — nearly half the countries on the continent. Thirteen of the 34 are exporters of fuel minerals, or had manufactured exports worth US$100 million or more in 1979, but only two of the 15 most-at-risk countries (scoring eight or nine) came into these categories.

Figure 8 lists 63 of the 64 critical countries, in order of the degree of criticality, and also in order of the severity of other problems faced. (Western Sahara, population ratio below 0.01, is not included because of insufficient data.)

The fact that a country is not included in the list does not necessarily mean that it is free of problems with regard to land resources and food supplies. As will be seen in the next chapter, an additional 14 countries would be critical for 1975, using low inputs, if a one-third allowance were made for non-food and food consumption factors, while another 19 might be critical as regards the land they actually cultivate. However, the land resources of these countries would be more adequate, in relation to their populations, than in the 64 critical countries.

Each country must assess for itself the extent to which it will or will not be able to import its potential food deficit. A combination of measures in agricultural development, population policy and economic development will be needed to avoid potential conflicts between population and land resources which are bound to have serious human, ecological and political consequences.

4. The present predicament

Alarming though the results quoted so far are, they fail to convey the full seriousness of the real situation.

The length-of-growing-period zone results and the country results are, essentially, an indication of the maximum numbers that could be sustained, at the level of a basic, largely vegetarian diet, from the *entire* resources of potentially cultivable land, if every hectare were used to produce food and all food was distributed equally among social groups. However, they tend to overstate the capacity of lands to support people at socially acceptable levels of diet, given existing inequalities, and with proper provision for forestry, mixed arable/livestock farming, fruit and vegetables and so on. Also, they do not, without adjustment, tell us the potential population-supporting capacity of *actually* cultivated lands.

This chapter discusses these factors in more detail, and attempts to adjust the regional results to take some account of them.

Allowance for other food and non-food needs

In the real world, it is extremely unlikely that the entire potentially cultivable area will be used to grow nothing but staple foods. Even for mere subsistence, land is required to grow vegetables and fruit to provide the vitamins and minerals essential for health. Land is also needed to grow fibre crops and trees to provide fuelwood and meet other needs. In addition, non-food cash crops may be grown to earn foreign exchange, or to provide cash for farmers to buy essential manufactured items such as pots and pans, items of furniture, bicycles or carts.

It is impossible to fix a precise figure for these non-staple requirements. One indicator comes from household budgets. These show that, even in the poorest homes, non-food requirements take up, at the very least, 20-30 percent of household expenditure (Ref. 12).

Land-use figures provide more guidance. In 1980, an average of 13 percent of the harvested area of the 90 countries studied for *Agriculture: Toward 2000* was given over to non-food crops. The share ranged from 10 percent in the Near East and 11 percent in Africa and Asia to as high as 21 percent in Latin America. An average of 56 percent of the area was under cereals but, over the 1978-80 period, about one sixth of cereal production in developing countries was used as feed for livestock. Taking these two factors together, more than 20 percent of the harvested area of the developing countries was not being used to grow basic foodstuffs, although the proportion in the poorer countries would be lower (Ref. 13).

Another land-use requirement is forestry. The land resources study assumes forests to be restricted to areas unsuitable for crops or for grassland supporting livestock. In practice, forestry, integrated into agriculture and community development, will be an essential requirement for fuelwood, fodder, construction and furniture, as well as for soil conservation. The study's results for humid tropical countries imply (although, of course, they do not advocate) the clearance of vast areas of closed forest, but again, in practice, such countries may often prefer to leave the forest as forest, for environmental as well as economic reasons.

Fuelwood is one of the most essential products of forests and trees, and requirements vary enormously from as little as 0.2 cubic metre per person in parts of the Indo-Gangetic plain up to 1.9 cubic metres per person in hilly and cooler parts of rural Africa (Ref. 14).

Fuelwood productivity varies from as low as 0.1 cubic metre per hectare per year in semi-arid shrub formations to 8 cubic metres a year in efficient plantations (Ref. 15). Thus, anything from 0.025 hectare to two or more hectares per person might be required to grow the fuelwood requirements.

Fuel needs, however, can be met in many different ways: by an increased use of fossil fuels, of biogas, of wind, water or solar energy, although it seems likely that dependence on fuelwood in developing countries will remain high for the foreseeable future. Fuelwood needs can be met by charcoal transported from closed forests, or by trees grown on field and garden fringes and roadsides, or on fallow land, where they would take up little in the way of potential arable land. Indeed, the use of selected nitrogen-fixing trees such as leucaena or acacia can result in increased agricultural productivity. If land for trees is not provided, the result can only be more serious land degradation.

Overall, then, it is not possible (outside of more detailed national surveys) to assign a precise figure for the land to be set aside for forestry, nor for the effect of this on food production and population-carrying capacity.

If fisheries were taken into account, the results might be improved for certain countries and regions. In 1977, fisheries contributed only 0.7 percent of calories to the average diet for developing market economies. As the rate of growth in fish output is not expected to keep pace with population, this contribution may even decline. Considered in terms of calories, therefore, fisheries would make very little difference to the population-supporting capacities.

However, fish made up a much higher proportion of protein intakes — averaging 4.4 percent in developing countries in 1977. If this were assessed toward meeting the minimum protein requirements at intermediate and high-input levels, it would allow a greater proportion of low-protein, high-calorie crops such as cassava to be grown. This would increase the carrying capacity. The effect could be significant in some countries, especially islands such as Indonesia, the Caribbean islands, and the Philippines (where fish accounted for 21 percent of protein intakes in 1977) and in countries with sizeable coastal or inland fisheries such as Ghana, Peru, Chad or Uganda (Ref. 17).

Other factors that could *increase* the carrying capacity would be major land improvements — such as terracing, drainage, flood control, etc., the development of specialized crop varieties adapted to less favourable conditions, or the reduction of waste *below* 10 percent of production.

Allowing for food consumption factors

The above considerations affect the amount of food the land could produce. There are other factors that tend to increase the food requirements of populations, in the long term, above the level assumed in the study — and thus to reduce carrying capacities even further.

The land resources study calculated potential population-supporting capacities by dividing the total calorie yield of the land by the average calorie intakes recommended by FAO/WHO for each country. This procedure is based on the assumption that food is equally distributed among social groups. In most developing countries (and indeed in developed countries), this is far from being the case. Poorer income groups, families with smallholdings or with no land, women and children often have calorie intakes below their average requirements. Widespread malnutrition can exist side by side with adequate *average* calorie intakes for a country or region.

Because of the considerable inequalities in food distribution in many developing countries, the Fourth World Food Survey calculated that to be sure the proportion of malnourished fell below 5 percent, the average calorie supply in most developing countries would have to reach 3 000 calories per day — 36 percent more than the actual intake in 1972-74. To allow for this, the population-supporting results would need to be reduced accordingly (Ref. 18).

However, increasing the average calorie intake is not a very speedy or efficient strategy for reducing the numbers of malnourished, as it implies considerable waste among groups consuming far more than their requirements. An alternative is to set up supplementary feeding, food-for-work and food subsidy programmes for them. For this reason, no uniform increase in calorie requirements to allow for inequality can be assumed.

Further considerations are likely to increase average calorie requirements and demands above those used for the present study. As population growth slows down and the proportion of adults increases in developing countries, from the present 62 percent toward the 80 percent of Western Europe, the average calorie requirements for each country will rise, and the population-supporting capacities of the land decline accordingly.

As incomes rise, people tend to consume more calories than they need, and to consume more meat and dairy products that use up a lot of crop production as feed. In 1978-80, the developed countries' average calorie intakes were 15 to 50 percent higher than their requirements (Ref. 18). The population that land resources could feed at these levels of intake would, therefore, be 13 to 33 percent lower than those assessed in the present study. The proportion of cereals used as feed also increases — in the developed countries, it was no less than 61 percent in 1975-79 (Ref. 19).

The one-third deduction

The complexity of all these qualifying factors, and the fact that several of them change over time and vary from country to country make it impossible to indicate an accurate figure, consistent for all regions and time periods, by which the supporting capacities could be reduced to give a more realistic picture.

However, an overall deduction of one third seems a moderate, even conservative, allowance. The study's figures for supporting capacity, when reduced by this amount, would indicate the populations that could be sustained at an acceptable although still fairly modest level. They would allow some leeway for non-food crops, forestry, changing age structures and demands for calories and protein, and some degree of inequality (see Table 2).

Table 2. POPULATION RATIOS
(entire land base, allowing one-third deduction for non-food crops, inequality, impact of development)

	INPUT LEVEL	AFRICA	SW ASIA	SE ASIA	CENTRAL AMERICA	SOUTH AMERICA
1975	Low	1.96	0.53	0.75	1.09	3.93
	Intermediate	7.70	0.84	2.01	2.08	15.93
	High	22.50	1.31	3.39	7.64	38.11
2000	Low	1.07	0.45	0.84	0.90	2.33
	Intermediate	3.08	0.60	1.50	1.71	8.87
	High	10.90	0.82	2.18	4.01	20.99
2025	Low	0.65	0.50	1.34	1.06	1.77
	Intermediate	2.09	0.55	1.55	1.49	6.32
	High	5.88	0.65	1.95	2.99	14.75

The ratio indicates the present or projected population which the land resources can feed. Ratios below 1 are considered to be critical.

This adjustment makes a considerable difference to the results, often swinging the balance of the land resource from a sufficiency to a deficiency.

The number of critical countries with low inputs in the year 2000, for example, would increase from 64 to 75, with four additional countries in Southeast Asia, four more in Africa, two in Central America and one in Southwest Asia. The number of critical countries with intermediate inputs would rise from 36 to 43, and with high inputs, from 19 to 29.

The regional results are presented in Table 3. This gives the population-supporting ratios — the number of times the present or projected population that the land resource can support. Obviously, the one-third deduction depresses the ratios everywhere, but it also makes several regions critical that were not critical *without* the reduction. For the year 1975, Southeast Asia becomes critical with the low-input level, and Southwest Asia even with intermediate inputs. The impact on the year 2000 results is even greater. Southeast Asia and Central America become critical with low inputs, in addition to Southwest Asia, while Africa has a bare sufficiency. Southwest Asia is now critical even with high inputs.

Table 3 also attempts to assess the possible situation for the year 2025. United Nations projections so far ahead were not available when the basic research on the Land Resources for Future Populations Project was done. The year 2025 time frame, although it may seem distant, is important in that the jump in human numbers between 2000 and 2025 will be as large as that between 1975 and 2000. The year 2025 regional results presented here are based on the UN's medium population projection (Ref. 20). They also assume that the potential of the land resource will increase at the same rate as that projected for 1975-2000, mainly owing to the expansion of irrigation.

In regions where irrigation is expected to expand significantly, there is an improvement in the balance between land resources and population — Southeast Asia and Central America cease to be critical in 2025 at the low level of inputs, and the ratio for Southwest Asia improves slightly — although it must be emphasized that these improvements depend on achieving sustained expansion of irrigation and on continued progress in reducing the population fertility level. For Africa and South America, where irrigation is projected to grow more slowly than population, the situation deteriorates; indeed Africa, where population growth will be fastest, would become critical by 2025 with low inputs.

Actually cultivated land

The land resources study set out to assess the potential for food production of *all* the land resources that could possibly be used for rain-fed agriculture.

However, it is also useful to ask the question: what is the population-supporting capacity of the land that is *actually cultivated*, and how does it compare with populations? This gives a better idea of the possible extent of overloading of currently

Table 3. POTENTIAL OF ACTUALLY CULTIVATED LAND
(with one-third deduction)

INPUT LEVEL	AFRICA	SW ASIA	SE ASIA	CENTRAL AMERICA	SOUTH AMERICA
		YEAR 1975			
Low	0.8	0.5	0.7	0.5	0.6
Intermediate	3.3	0.8	1.85	1.4	2.4
High	9.6	1.3	3.1	3.7	5.8
		YEAR 2000			
Low	0.55	0.45	0.85	0.5	0.5
Intermediate	2.0	0.60	1.50	0.9	1.8
High	5.7	0.80	2.20	2.2	4.3
		YEAR 2025			
Low	0.4	0.5	1.35	0.7	0.5
Intermediate	1.3	0.55	1.55	0.9	1.7
High	3.7	0.65	1.95	1.9	4.0

Notes:
- Figures show the fractions of expected populations that regional lands can support. Ratios below 1 are considered to be critical.
- Africa's cultivated land is assumed to be double the figure quoted in the *FAO Production Yearbook* to allow for fallow land.
- Southwest Asia is already cultivating more land than is suitable. Therefore, no further expansion is assumed and the unadjusted study results are used, less one third.
- Southeast Asia was using 92 percent of its suitable land in 1975. At 1969-71 rates, the region would reach 100 percent before 2000. Therefore, the results for 2000 and 2025 are the unadjusted results, less one third.
- The results for the other regions assume that the cultivated area will expand at the same rate between 1975 and 2000 and between 2000 and 2025 as applied between 1969-71 and 1980.
- The year 2025 results assumed that potential continues to increase at the same rate as between 1975 and 2000. This involves primarily expansion of irrigation.

used land. The question can also be asked: what land seems, on present trends, likely to be occupied by 2000 and 2025? The reply will give an indication of the realistic prospects for food production and of the progress that may be needed in expanding the cultivated area, increasing inputs to agriculture or reducing fertility, in order to bring the potential of the land and the food needs of expected populations more into balance where foreign-exchange sources are inadequate or insecure.

The land resources of developing countries are very unevenly distributed between regions and between countries (see Fig. 9, Land reserves and cultivated area, 1975-2025).

Southwest Asia, in 1975, was already cultivating an area larger than its potentially cultivable area — some of the excess was irrigated land, some unsuitable land — and therefore has little or no room for further expansion.

Southeast Asia is also approaching the limits of its suitable land: 92 percent of the potential was under cultivation in 1975. Most of the land reserves that exist are in Indonesia, Malaysia, Burma, Laos and Democratic Kampuchea, each of which is cultivating less than half of its potentially suitable area (Ref. 21).

Central America still has some room for expansion. In 1975, just under half the potentially suitable area was cultivated. Here, most of the islands were closer to their limits, while most of the mainland countries (with the exception of El Salvador) had reserves, mostly of lowland and forest.

According to some official figures, Africa was cultivating only just over a fifth of its potential cropland in 1975, although the inclusion of fallow land would increase this proportion considerably. Most of the reserves were concentrated in central Africa. The countries here, along with Angola, Mozambique, the United Republic of Tanzania, Zambia and Zimbabwe, were cultivating a fifth or less of their potential. In contrast, the countries of North Africa, the Sahel and East Africa were running short of suitable land, in many cases cultivating more than 80 percent of the potential arable area.

South America is using the smallest proportion of its potential of any region — only 15 percent in 1975. Most countries have enormous reserves of suitable land. Only three countries were cultivating more than a fifth of their area in 1975: Ecuador (44 percent), Argentina (55 percent) and Chile (85.5 percent) (Ref. 22).

The cultivated area has not been expanding as rapidly as population in many regions. In South America, the area under arable and permanent crops grew by 1.2 percent a year between 1969-71 and 1980. In Africa, the rate of expansion was 0.78 percent per year, and in Central America, 0.5 percent. Southeast Asia's cropland grew by an annual 0.46 percent, while in the Near East the annual growth rate was 0.26 percent (Ref. 23). Figure 9 shows the areas that would be occupied by 2000 and 2025 if these growth rates continued.

It is not known whether the remaining land reserves are of the same productivity as currently utilized land. There is evidence to suggest that the productivity of the reserves may be higher, but, for the sake of simplicity, it is assumed that the potential productivity of the unused land is the same as that of the land under cultivation.

The figures on actual and projected land use can be used to adjust the regional results of the study. In 1975, for example, South America was cultivating only 15 percent of the potentially suitable land; so the potential population-supporting capacity of the entire land resource is reduced by 85 percent to find the supporting capacity of the land under cultivation in 1975.

Table 3 presents the regional results for the population-supporting capacity of *actually* cultivated land, including the one-third deduction for non-food and food consumption factors. These modified results, therefore, give an idea of the populations that the areas at present, or likely to be, cultivated can sustain to provide realistic standards of living.

These adjusted results must give grave cause for concern. As can be seen, every region in the developing world emerges as unable to feed its 1975 population from its cultivated lands using low inputs. By the year 2000, they would still all be critical with low inputs, and the situation will have deteriorated for every region except Southeast Asia, which will have the slowest population growth and the largest expansion of irrigation.

FIGURE 9

LAND RESERVES AND CULTIVATED AREA
(million ha)

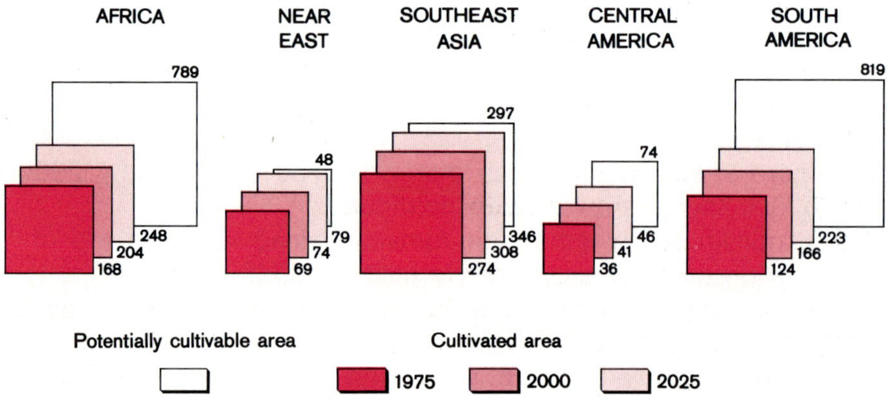

Central America would be critical with intermediate inputs and Southwest Asia with high inputs.

By the year 2025, the low input results will have improved for Southeast Asia, and slightly for Southwest Asia, Central and South America, mainly because population growth rates will have slowed while irrigated production is assumed to continue growing as before. If irrigation does not continue to expand at the projected rates for 1975-2000, however, this improvement in carrying capacity may not take place. However, in Africa, the population will be even more in excess of the supporting capacity with the low level of inputs.

A similar adjustment for the proportion of the area that is cultivated can be made on a national basis. This has a more dramatic effect than the one-third deduction. Out of 53 countries which were not critical in 1975 in relation to their entire land resources, and for which data are available from the *Agriculture: Toward 2000* study, no less than 26 became critical in relation to their actually cultivated land. These countries do, however, have one major advantage over the countries that emerge as critical on their *entire* lands: they can at least look to the expansion of their cultivated area as one way of relieving the situation.

The total number of countries whose land resources were unable to feed their populations rises considerably when *both* the modifying factors — one-third deduction and adjustment for cultivated land — are taken into account. Map 3 shows the additional countries that become critical as each adjustment is made. Altogether, the total number of critical countries with low inputs in 1975 rose from 54 out of 117 without adjustments to 68 with the one-third deduction, 87 with respect to cultivated land, and no less than 99 on their cultivated land less one third — 85 percent of the total. One indicator that this offers a realistic assessment of the potential of the land in use is that, out of the 90 countries studied for *Agriculture: Toward 2000*, 72 — or 80 percent — were not self-sufficient in basic foods in 1978-79 (Ref. 21).

The fact that, neverthless, many of the countries and some of the regions were more self-sufficient than the adjusted low-input results suggest, can be accounted for by a number of factors. First, all regions except Africa are, on average, some way above the low level of inputs. Second, in many countries, and in the African and Southeast Asian regions, average calorie intakes are below the requirements assumed for the land resources study — thus, higher populations are being sustained but at a lower level of health and physical energy.

5. The outlook: potentials and pitfalls

As can be seen in the light of the foregoing analysis, the real situation that confronts many countries and regions is extremely serious.

This chapter will consider in detail the trends that most closely affect the outlook for the future: the prospects for population growth and for progress in agriculture.

Population

Concern about the population explosion tended to diminish after the world food crisis of 1973-74 as problems such as the world recession and world security came to the fore.

However, the most critical phase of the population explosion, and its most serious potential confrontation with land, water and energy resources in many areas are perhaps still to come. The years from 1980 to 2020 will see the largest annual additions to human numbers in history, and present actions, or failures to act, will determine whether these numbers can be accommodated without widespread famine, ecological catastrophe or strife.

On the UN medium projection, the world's population is expected to grow from 4.43 thousand million in 1980 to 7.81 thousand million in 2020. No less than 93 percent of this growth will occur in the developing countries, whose populations are projected to grow from 3.3 thousand million to 6.45 thousand million over the same period. Although the average *rate* of growth in developing countries as a whole has in fact been slowing down since 1965-70, the absolute numbers added each year have been rising, and will continue to rise. Between 1960 and 1980, an average of 60 million a year were added; in the 1980s, the annual increase will be 73 million a year, and will rise to more than 80 million a year between 1990 and 2010 (Ref. 20).

One of the most disturbing features of this growth pattern is that it will be fastest precisely in those regions where land resources are least adequate to meet the needs of their populations. Around the year 2000, the slowest population growth rates of the regions studied will be in South and Southeast Asia, where growth will average below 1.5 percent a year, and in tropical South America, where it will average just over 2 percent a year.

The fastest growth rates will be found in the regions with most countries that appear critical in the study for the year 2000: in Middle America, Africa and Southwest Asia.[1]

In Middle America, the population was growing at 2.88 percent a year in 1980-85. By 2000-05, it will be still growing at 2.13 percent a year, although it is projected to slow down fairly rapidly after that. Nevertheless, the population of the study's Central America region is projected to rise from 123 million in 1980 to 304 million in 2025, two and a half times the 1980 level.

Growth rates in Southwest Asia will be very similar to these, slowing down from 2.89 percent a year in 1980-85 to 2.18 percent in 2000-05. Here, the population will rise from 98 million in 1980 to 264 million in 2025 — 2.7 times the 1980 figure.

The most rapid growth is expected in Africa, whose 1980 population of 470 million (including southern Africa) is projected to grow to 853 million by the end of the century and to no less than 1 542 million by 2025.

Death rates have fallen in Africa and there have been few signs of population fertility decline in most African countries. As

---

[1] The regions quoted in this section are those of the UN population projections and do not always coincide with those in the land resources study; — they are the UN Population Division's regions.

FIGURE 10

POPULATION GROWTH 1950-2100
(millions)

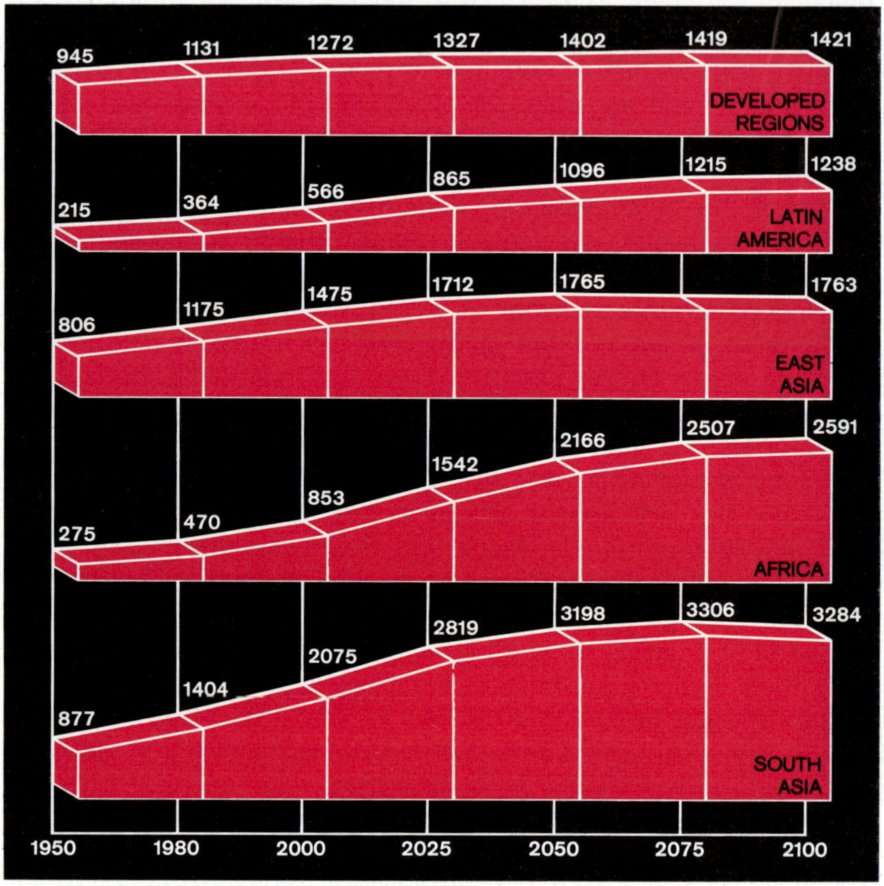

SOURCE: POPULATION BULLETIN OF THE UNITED NATIONS, No 14, 1982, (ST/ESA/SER N/14)

a result, the population is growing faster than in any other region, at 3 percent a year from 1980 to 1985, and is projected to be still expanding at 1.91 percent a year 40 years from now. Within Africa, East and West Africa, the location of some of the most critical countries in the region, are expected to grow fastest. The year 2025 population of Kenya, for example, is projected to be more than six times its 1980 population, that of Rwanda almost five times, and of Nigeria and Uganda, around four and a half times (see Appendix 2 for more examples).

Agricultural development

Given these rates of population growth, can the expansion of potential food production keep pace?

Expansion of the land base has already been considered and taken into account in the projections for the years 2000 and 2025. As can be seen, the cultivated area is not expected to expand in line with the population in any region. Thus, the average arable area per person is expected to decline, in the 90 countries studied for *Agriculture: Toward 2000*, from 0.37 hectare per person in 1975 to 0.25 hectare in 2000 (Ref. 6).

Obstacles to expansion

Indeed, it is possible that the rate of expansion may slow down. The bulk of the high potential land reserves that remain in developing countries are found in the humid tropics — in South America, central Africa and parts of Southeast Asia. These remaining areas are for the most part more difficult to settle and farm than existing cultivated areas. They are often under tropical forest, laborious to clear, and lacking in infrastructure of any kind. Pests and diseases of humans, livestock and plants are more common. Topsoils are often thin and require careful management if land degradation is to be avoided. Unplanned or badly planned settlement — for example, allowing cattle-ranching in cleared tropical forests — has often resulted in ecological mishaps. The general lesson of even carefully planned settlement is that it can often be a slow and expensive business.

Thus, there may be serious obstacles to the rapid exploitation of the bulk of land reserves in these areas, many of which indeed may never be used for crop production. Increasing urbanization will reduce the number of potential settlers, and many of the countries with large tropical forest areas may wish to keep these areas for forestry use.

The degradation threat

Meanwhile, the land that is already in use for food production is threatened by land degradation. The land resources study showed the large potential losses of land and productivity that could result from unchecked degradation. The United Nations Desertification Map of the World indicated that 43 percent of the non-desert area of Africa was at risk of desertification. The equivalent proportion for Asia was 32 percent, and for South America 19 percent (Ref. 24).

It is harder to assess what is actually happening on the ground to quantify to what extent these risks are becoming reality. No systematic global survey has been made, but regional surveys give some indication. In Africa north of the equator, it was found that 35 percent of the total area is affected by wind or water erosion or salinization. In the Near East, the proportion of the area affected is 65 percent, and in India, 51 percent of the total area is thought to be affected by erosion, salinity or flooding (Refs 25 and 26).

The growing scarcity of fuelwood is another pointer to the extent of environmental degradation. FAO's Forestry Department studied the situation in 95 developing countries. Altogether, 69 of these faced problems with fuelwood supplies; 42 had a deficit situation in part or all of their territory, and could meet their fuelwood needs only by depleting tree stocks. Another 27 countries faced an acute scarcity, and even overcutting could not supply their needs; 34 of the countries with deficit and acute scarcity situations were in Africa (Refs 15 and 16).

There is a disturbing degree of correspondence between the areas at risk of desertification and deficient in fuelwood and those areas highlighted by the present study as having inadequate land resources to feed their populations with low inputs. This can be seen clearly on Map 4.

The reason for this connection is partly climatic: the critical zones are often cooler areas. Cool areas, in the tropics, are invariably more hilly and, therefore, more liable to erosion. Part of the reason is economic and ecological: most of the farmers in these areas are operating with low inputs which cannot provide even a livelihood. The sheer necessity of survival forces them to overexploit the land, farming more marginal, steeply sloping or arid areas, reducing fallow periods below the levels needed to maintain soil fertility and soil structure. Thus, the inadequacy of the land resource to feed its existing population at prevailing levels of technology can become a principal cause of land degradation.

Areas of particular pressure are the Andes and Northeast Brazil, El Salvador, Guatemala and Haiti; the Sahelian countries, the Horn of Africa and East Africa; much of Southwest Asia and Afghanistan; and parts of South Asia, Java and the Philippines. In several of these areas — especially Latin America — inequality in landholdings is a major reason why small peasants are forced to farm marginal areas.

There are always other complex political and social factors involved, which can result in disproportionate shares of often limited national budgets being spent on swords rather than on ploughshares, thus compounding the dangers and delaying solutions.

Resolving the potential conflict between land resources, inputs, populations and economic development, therefore, becomes of great importance to world peace, as well as to human welfare and to the conservation of the environment.

Progress to higher input levels

The shortage of potentially cultivable land in land-scarce countries, and the relative slowness of settlement of new land in land-abundant countries mean that increasing the level of inputs to agriculture will be vital to achieving a better balance between populations and land resources.

There seems to be no foreseeable reason, in terms of supplies of inputs, why the majority of developing countries should not be able to reach the higher input levels. The mineral fertilizer requirements of the intermediate level of inputs, for example, would amount to about 75 kg of nutrients per hectare per year — roughly the level attained by Sri Lanka in 1980. The high level would imply around 150 kg per hectare, attained by Costa Rica. (Lower quality land would require more; better quality land, less.)

The total effort involved in attaining these levels would be considerable: for example, if the intermediate input level were applied to all suitable potentially cultivable cropland in the 117 countries, this would require no less than 152 million tonnes of nutrients, which would cost, at 1983 prices, roughly US$22 thousand million. The high level would involve double these quantities. Against this, developing countries' fertilizer consumption in 1981/82 was only 38 million tonnes. Nevertheless, there is no reason to assume that the world could not produce, or many developing countries afford, the required amounts, given sufficient investment in agriculture.

The gradual depletion of the present feed stocks for commercial nitrogenous fertilizer production — oil and natural gas — may raise the price of commercial fertilizers, but this will provide a greater stimulus for the more efficient use of commercial fertilizers, and for increased use of organic fertilizers and biologically fixed nitrogen in carefully planned integrated plant nutrition systems.

The most serious obstacles will be economic and human: the ability of poorer countries and farmers to pay for increased inputs and to organize the changes that will be needed to ensure efficient use of land and inputs.

The adjusted results presented in Chapter 4 (Table 3) suggest a very rough timetable, to indicate what input level each of the regions should reach by certain dates if they are to be able to feed their own populations from their own lands. This can then be compared with the rates of progress that have been achieved over recent years, as illustrated in Table 4, Progress on yields.

Southwest Asia would need to attain high levels of inputs as soon as possible, and at any rate before 2000. The region is, on average, currently only 40 percent of the way between low and intermediate levels. If wheat yields continued to increase each year by the same amount as between 1969-71 and 1979-81, the region would attain the intermediate-level equivalent by 2000 and the high level perhaps 40 years later. If it could maintain the same *rate* of improvement in yield, it could reach the respective levels in 1995 and 2010. Thus, strenuous efforts in agriculture might enable the region to avoid excessive dependence on food imports, but a more rapid decline in population fertility would help, and careful planning of sources of foreign exchange will be advisable, especially for Afghanistan, Democratic Yemen, Yemen and Jordan (see Appendix 2).

Southeast Asia as a whole made the kind of progress in the 1970s that would enable it to maintain its self-sufficiency in food over the long term. Yields have improved significantly over the past decade: if the same annual *increment* were achieved, the intermediate level would be reached before 2010 for rice, while the high level would be reached by 2060. Keeping up the same annual *rate* of increase would enable the intermediate level to be reached by 2000 and the high level by 2030. Thus, continuation of present trends in agricultural development and population fertility decline would allow the region as a whole to avoid any increase in dependence on imports. On the other hand, the margins are modest and leave little room for unexpectedly high rates of population fertility or slacking off of efforts in agriculture. Certain countries, most notably Viet Nam and Bangladesh, will need to intensify efforts in the population and agriculture fields.

Latin America is at present, on its currently cultivated lands, about 30 percent above the low level of inputs, which is sufficient to remove the region from the critical list. Yields of maize, which occupies three times the area of wheat in the region, have improved quite rapidly, and would reach the equivalent of intermediate level between 2010 and 2025 and of high level between 2020 and 2070 if the progress rates of the 1970s were maintained.

In South America, few if any countries seem likely to run into serious conflicts between land resources and populations at *national* level, but subnational problems are already manifest in the Andes and Northeast Brazil which call for particular attention.

The situation in Central America and the Caribbean is more restrictive. If this region were to feed itself from its cultivated area, it would need to have passed beyond the intermediate level of inputs by the year 2000. On present trends, that would not be achieved. Thus, increased efforts in agricultural development would be required, while accelerated reduction in

Table 4. PROGRESS ON YIELDS

REGION and CROP	YIELD in tonnes per hectare 1969-71	1979-81	INCREASE PER YEAR kg/ha	percent	DATE[1] input level intermediate	high
AFRICA						
sorghum	0.66	0.67	1.0	0.15	—	—
maize	1.00	0.95	—	—	—	—
millet	0.58	0.59	—	0.05	—	—
cassava	6.64	6.37	—	—	—	—
NEAR EAST						
wheat	1.07	1.46	39.0	—	1995-2000	2010-2040
FAR EAST						
wheat	1.2	1.55	35.0	—	1995*2000	2015-2040
rice	1.84	2.19	35.0	—	2000-2010	2030-2060
LATIN AMERICA						
wheat	1.39	1.49	10.0	—	2040-2060	2110-2200
maize	1.43	1.82	39.0	—	2010-2020	2025-2070

Source: FAO PRODUCTION YEARBOOK, 1981.

[1] In the last two columns, the first date of each pair indicates the year in which the stated level of input would be reached if the increase in kg/ha were maintained. The second date shows when the level would be reached if the percentage *rate* of increase were maintained.

population fertility would help, most crucially in El Salvador and Haiti but also in Guatemala.

It is Africa, however, that faces the greatest challenge. Currently operating roughly at the low level of inputs, it would need to progress something like half way toward the intermediate level by 2000; in other words, to be close to the present Near Eastern level. This would be by no means impossible, given a concerted national and international effort, but present trends are not encouraging. Over the 1970s, yields of the most important African food crops either stagnated or declined, despite some progress in the use of fertilizers. Moreover, many African countries have not yet developed stable sources of foreign exchange that might cover their potential deficits. On present trends, much of Africa — especially the Sahel, the dry belt in the south of the continent, the Horn of Africa and East Africa — appears to be heading toward human and ecological tragedy, unless far greater priority is given to agricultural development, conservation, and programmes to reduce population fertility than has hitherto been the case.

6. What can be done?

The picture that emerges from the land resources project, at country level, is a very diverse one.

There are a number of sparsely populated tropical countries that are capable of feeding many times their expected populations from their own lands, even with low inputs. At the other extreme, there will be, by year 2000, 17 countries which will not be able to feed their populations from their own lands unless they rise above the intermediate level of inputs, and another 19 which will be unable to do so, even with high inputs.

There are countries with vast land reserves, others with almost none; countries with still very rapid population growth rates, others that are showing encouraging signs of a slowing down in such rates; countries with ample and secure sources of foreign exchange with which to buy any food imports they need, others with inadequate or fluctuating sources.

It is, thus, impossible to outline general prescriptions applicable to all countries: the policy measures required will be as diverse as the country situations. In all countries, the task will be to ensure that a satisfactory balance can be achieved between the food needs of future populations, the production possibilities of the land resource, and the foreign-exchange earning capacity of the economy. In most countries, this will require a carefully coordinated combination of policies, stretching from agricultural development and agrarian reform to policies and programmes on the size and distribution of populations, and plans and provisions for stable sources of foreign exchange to buy food imports where these are required.

The international dimension

Some of these solutions will involve international cooperation, either at global or at regional level.

First, it is essential that developing countries enjoy easy access to essential technology.

The efforts of the international research networks should continue to focus increasingly on crops suited to local conditions, and on the constraints that stand in the way of higher yields.

In cases where countries cannot realistically achieve food self-sufficiency, or where this is not economically desirable, the prevailing international economic system does not always enable them to finance their imports of food or farming inputs on a stable basis. Such countries will require reliable earnings from exports of primary commodities as well as of processed and partly processed commodities and manufactures. The fluctuating prices of many commodities and the recently rising levels of *de facto* protectionism make this difficult. Thus, action to reduce protectionist barriers is needed, as well as progress toward new arrangements for commodities that ensure more stable and equitable prices.

There is also scope for South-South initiatives. A strong case can be made out, in terms of political independence, food security and regional economic development, for a greater measure of *regional* self-sufficiency in food.

The potential for regional food self-sufficiency is good in four of the five regions. Most deficit countries find themselves adjacent, or at least close, to countries with a potential surplus within the same region.

For example, the 22 African countries that were critical with low inputs in 1975 could all meet their expected food deficits by importing food from neighbouring African surplus countries. These 22 deficit countries required extra supplies for 61 million people, while ten adjacent surplus countries were capable of feeding almost 550 million people in excess of their own populations. Thus, a mere 11 percent of the potential surplus in these countries would suffice to meet the deficit needs of the critical countries. With higher levels of inputs, the proportion of the surplus required to meet the deficit is much smaller. The beneficial effects of such arrangements on intraregional trade, communication and cooperation are obvious.

There is, however, a problem as to how the potential surpluses could actually be produced. Many of the potential surplus countries are sparsely populated. There is also the problem of how the deficit countries might finance the necessary imports. One possible solution could be, within regions, for migrants from deficit countries to supply, perhaps temporarily, some of the labour to produce extra food from the underused high potential land in the surplus countries, while their remittances to their home countries would help the latter to pay for their food imports. However, without a planned solution to this problem, international tensions could easily arise between acutely overcrowded deficit areas and neighbouring areas with large unused potential land reserves — if only from spontaneous migrations.

National dimensions

The land resource study's results, among other things, indicate the potential for national self-sufficiency in food. The study does not propose or imply that self-sufficiency must be a goal for every country. There are some countries that cannot realistically hope to achieve self-sufficiency. There are others where it may make better economic sense, in terms of comparative advantage, to use part of the suitable land to grow valuable export crops that can earn more, in foreign exchange, and import the food crops that the same land might have grown.

Where self-sufficiency is not the goal, planning for stable sources of foreign exchange becomes important. These may include exports of high-value cash crops, minerals, manufactures, or tourism.

Nevertheless, there will inevitably be a large number of

FIGURE 11

THE POTENTIAL FOR LAND REFORM

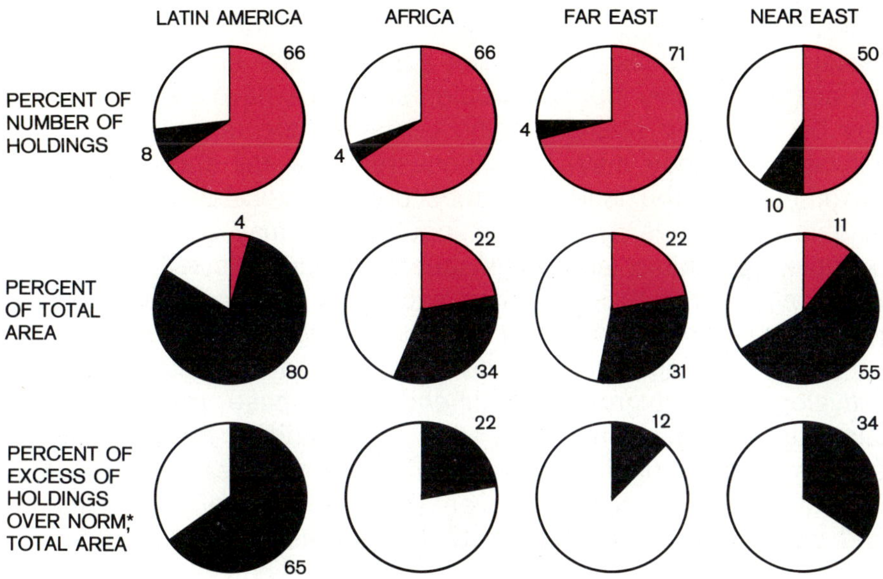

DEFINITIONS

	FAR EAST AND AFRICA	NEAR EAST	LATIN AMERICA
Small holdings:	below 2 ha	5 ha	10 ha
Large holdings:	above 10 ha	20 ha	100 ha

* Norm for large holdings is: 10 ha in Africa and Far East, 20 ha in Near East, 100 ha in Latin America

countries that wish to maintain or increase their self-sufficiency in food as a contribution to increased food security. In such cases, the most important policies will concern the balance between the possibilities of national food production and the future levels of populations.

Countries that show up as critical for a certain date, at the level of inputs they are closest to using, may well wish to raise their inputs closer to the level that would remove them from the critical list. Countries that do not appear to be critical, however, are not necessarily free of problems: as we have seen, they may be critical in relation to the land they are actually cultivating, or they may, for various reasons, fall short even of the output that the land resources project indicates they should be able to achieve, given their inputs. This may occur, for example, if suitable land within operating farms is left uncultivated, or if the marketing and price structure does not give the farmer the incentive to produce additional output.

These countries that appear as critical on their *currently cultivated* land, but not on their *entire* land base, may be able to calculate what feasible combination of expanding the cultivated area and increasing inputs will enable them to reach the level of self-sufficiency desired. However, the difficulties of successful land settlement should not be underestimated. The humid tropical areas have the greatest scope for expansion: their potential with suitable crops such as rice, cassava or oil palm is high, but their ecological sensitivity is also high, and settlement demands very careful attention to soil management and tree and shrub cover. The reasons why human populations have tended to avoid such areas must also be tackled, with a proper infrastructure of transport, health and other services and solutions for endemic problems such as tsetse fly.

In countries with scarce land reserves, intensifying the use of existing land will be of crucial importance. This will mean increasing the level of material inputs and higher use of fertilizers together with improved seeds and pesticides made available with the necessary back-up of research and extension services to smallholders and farmers in less favoured areas as well as to large landowners and prime agricultural zones. It will also mean redesigning marketing and pricing systems to make sure farmers have an incentive to increase their output, and improving administrative capacities to manage agricultural and rural development. In many countries, this may mean increasing the share of the budget devoted to rural areas as opposed to urban areas.

Of course, higher inputs need not equate with Western farming methods. The solutions will need to be suited to local ecological and social conditions. They may involve partially mechanized labour-intensive farms which, as the experience of labour-intensive smallholding in Hungary or Japan demonstrates, can produce just as good yields per hectare as complete mechanization. Such solutions may involve, at present, unconventional techniques such as minimum tillage, the use of "living mulches" of ground-level legume crops, or "alley cropping" — growing crops between rows of legume trees like leucaena. They must, in this era of increasingly expensive energy, aim to be energy-efficient, using chemical fertilizer to maximum effect, in combination with as much organic material as can be recycled by way of compost, mulches, biogas, etc., and using biological nitrogen fixation to maximum advantage.

The development of new seed varieties now in progress should help considerably. The first phase of the "green revolution" produced seeds that performed very well, but demanded optimum conditions and high inputs. In recent years, more emphasis has been placed on developing seeds that can do well in poor soils and adverse climates, which use fertilizer more efficiently, and which have in-built resistance to many pests and diseases — thus reducing the need for pesticides and fungicides.

Land improvements, on a large or small scale, can improve the results of the study, which assumed that very sloping land or flooded land, for example, could not be used to grow crops. Works such as terracing, levelling, reclamation, flood control and drainage, when properly maintained, would actually increase the amount of potentially suitable land available. Such works, when they use labour-intensive methods, would also help to create rural jobs in large numbers. Several countries have increased their land resources in this way — or reversed the degradation process. Since around 1960, for example, Pakistan has reduced

FIGURE 12

DEVELOPING COUNTRIES WITH LOW WATER AVAILABILITY
(1000 cubic metres per person, year 2000)

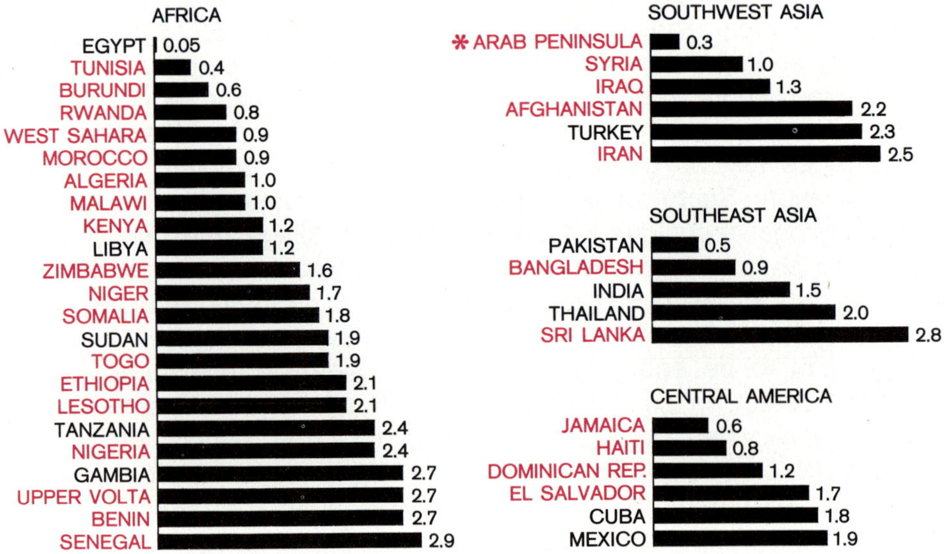

Countries in colour are critical for low inputs, 2000

* 8 Countries

SOURCE: GLOBAL 2000 REPORT TO THE PRESIDENT TECHNICAL REPORT PAGE 156-7

the area affected by salinization through improved drainage. More recently, massive food-for-work programmes, backed by food aid, have enabled Bangladesh to carry out far-reaching flood control measures, and Ethiopia to terrace steep slopes so that they can be cultivated without erosion risks.

Such enrichments of land resources to feed populations are, however, only attainable at considerable capital investment cost and do involve provision of additional funds on a recurrent basis to ensure necessary maintenance works. The economics of such land-improvement measures require careful study.

Another way in which the results of the study could be improved upon for some countries is expansion and intensification of irrigation. The land resources study was concerned with the potential of land for rain-fed agriculture: it took into account food production from irrigation, but at a *fixed* and fairly high level of inputs projected by the *Agriculture: Toward 2000* study. If inputs to irrigation were increased above this level, the potential contribution of irrigation would be higher. The same would apply if the irrigated area was expanded faster than projected. Figure 14 shows the present and projected extent of irrigation and rough estimates of the ultimate potential.

However, the ease of expanding irrigation should not be overestimated. The cost of fully equipping an area is very high, and it is highest where it is most urgently needed. In South Asia, costs start at US$1 600 per hectare. In Africa, they start at US$9 500 per hectare, raising problems not only of funding but sometimes even of economic viability (Ref. 27). The area that is *easily* irrigable is much smaller than the potential indicated in the figure: one study puts the African potential at only 18 million hectares, if only local water storage and transport are considered and only surplus from mountain areas is used (Ref. 28). Water availability will also be a growing problem in many areas: in general, except for a few very long rivers, the potential for low-cost irrigation expansion is greatest in the countries with the greatest potential for rain-fed agriculture. Those with lowest rain-fed potential also have the least potential for irrigation.

The per caput availability of water may be halved by the year 2000 in many developing countries, and expanding industry will take a growing proportion of supplies (Ref. 29). Some countries, as shown in Figure 12, will have very low water availability per person by 2000 — and these levels will again in many cases be *halved* by 2035. It is worth noting that, out of the 47 countries in this figure, no less than 36 will be critical with low inputs in 2000. Thus, water shortages will often compound the effects of inadequate land resources for food production — of which they are often one of the causes.

Improving agrarian structures can make a considerable contribution to achieving the production potentials that the land is theoretically capable of: existing arrangements can often depress production. For example, where tenancy arrangements are inequitable and rents or landlords' shares are excessive, tenants may lack the incentive to invest in capital improvements or to achieve higher yields.

The present distribution of landownership is illustrated in Figure 11. It is highly inequitable in many areas, and may often be a significant factor in depressing production (Ref. 30). Large farms have less incentive to use land intensively. Studies in nine Latin American countries, for example, showed that the proportion of the farm area cultivated in small farms was between two and a half and 16 times greater than on large farms: the proportion of the farm area cultivated in large farms was below 10 percent in six of the nine countries (Ref. 31).

Figure 11 also indicates the regional potential for land reform — the share of the total land area that would be available for redistribution if ceilings of 10 hectares were imposed in Africa and the Far East, 20 hectares in the Near East, and 100 hectares in Latin America. A detailed study of six countries found that a radical land reform — distributing farmland equally among all families in the rural labour force — could increase food production by 10 percent for Pakistan and by 28 percent for Peninsular Malaysia and Colombia (Ref. 31).

Changes in the proportions of crops grown could often produce worthwhile gains, even without increasing the level of other inputs. Frequently, for historical and cultural reasons, countries are not growing the crops that could produce the highest potential calorie-protein yields per hectare. In Africa, for example, nearly 12 million hectares were under sorghum in 1981 — around 8 percent of the total area of cropland (Ref. 10). Yet, the Agro-ecological Zones Project showed that, over most of the sorghum range, millet (in the drier areas) and maize (in the less dry) would give better yields.

The land resources study, with the higher input levels, was able to calculate the crop mix that would give the highest potential production. In Africa, for example, this would involve a considerable increase in the share of land under upland rice and soybeans, as well as a shift from sorghum to millet and maize.

In Southwest Asia, winter wheat is grown very widely in zones where rainfall is too variable for wheat's requirements. A major shift into the less vulnerable winter barley would produce higher output for the same input. Central America and South America would require significant increases in the share of land devoted to sweet potato, white potato and groundnuts. Southeast Asia is closer to the optimum mix, but would benefit from increases in the share of land under banana/plantain, soybean, maize, groundnut and sweet potato.

A preliminary study for the Land Resources for Future Populations Project, for African countries only, found that changing the crop mix alone could produce worthwhile gains, even if other factors held constant. For the year 2000, with low inputs and no conservation measures, the study suggested that shifting from the present to the optimum crop mix could raise the potential population-supporting capacity by 58 percent and cut the number of critical countries by four (Ref. 32).

Crop-mix changes are, of course, not a simple matter. They may involve changes in diet, farming practices, patterns of labour requirement and marketing; but the difficulties can be overstated. Sub-Saharan Africa, for example, has seen a fairly rapid and damaging shift in tastes away from locally produced cereals to wheat, which has to be imported because only a few cool areas are suitable for its cultivation. Given subsidies and backing from community nutrition programmes and extensions workers, more desirable shifts in tastes toward the more productive locally grown crops could be engineered. The study's detailed results at the agro-ecological cell level provide guidance as to what crop changes might be best for specific areas: more detailed national-level studies would elaborate these results.

The *importance of conservation measures* is also heavily underlined by the study's results. The potential loss of food production that could occur if land degradation were to continue unchecked has already been pointed out. Thus, the use of full conservation measures without changing the crop mix or increasing the level of inputs could increase long-term land productivity at the low-input level by 33 percent in Africa and Central America, by 26 percent in South America, by 14 percent in Southeast Asia and by 5 percent in Southwest Asia.

The preliminary African study mentioned above showed that

the results of full conservation measures could be very dramatic, especially in ecologically sensitive areas: the potential population-supporting capacity with low inputs was increased by 150 percent or more in 11 countries in the region. Full conservation measures alone would reduce by seven the number of African countries unable to feed their year 2000 populations from their own lands with low inputs (Ref. 32).

Conservation measures can include terracing, use of agro-forestry combinations, maintenance of soil cover and soil organic matter by means of mulches, living mulches, minimum tillage, and so on. Of course, introducing such measures requires a considerable amount of organization and extension work and changes in labour requirements. In many cases, lasting measures could be carried out by way of food-for-work programmes.

Improving the distribution and utilization of food

Sustained increases in food production are essential, but in most countries improved distribution will be needed to make sure that the increased output leads to real reductions in malnutrition.

As has been seen, the study's unadjusted results imply the equal distribution of food between different income groups. The *actual* population-supporting capacity in any country will depend partly on the degree of inequality in distribution of food: the more unequal the distribution, the lower the population that can be supported from a given level of production. Thus, countries that can improve the distribution of food will be helping to improve the balance between their populations and their land resources. Improving the distribution involves improving the access of poor groups to the land with which to produce their own food, or to jobs which will provide the incomes to enable them to buy sufficient food.

Much food is also wasted between harvest and dining-table, from storage pests and moulds, milling losses, and losses in nutritional value from cooking. Average losses in developing countries have been roughly estimated to average 10 percent for cereals and 20 percent for fruit, roots and vegetables (Ref. 33). The present study assumes an overall post-harvest loss of 10 percent. If post-harvest losses can be cut below 10 percent by improved storage and processing methods, then population-supporting capacities will be increased correspondingly.

The population dimension

There are two sides to the balance between the size of populations and the capacity of nations to feed them. So far, we have looked at the measures that might be taken to raise the output of the land or earnings of foreign exchange. However, governments can also take action to influence the distribution, growth rate, and size of future populations.

The land resources study, and the detailed national-level studies that will follow, should enable governments to identify which areas are most likely to be overloaded in future, and which areas have unexploited potential. Resettlement programmes can help to direct migrants into the high-potential areas.

For governments that succeed in lowering the rate of their country's population growth, there can be considerable benefits. Rapid population growth brings with it a range of problems, for families and for countries as a whole. Most surveys show that, on average, children from larger families are less well nourished and less healthy than those from small families. Countries whose populations are growing rapidly have to spend more of their often scarce capital on simply expanding the quantity of facilities — school places, health centres, work places — rather than on improving their quality. The working populations also have to support much larger non-working populations of young people than countries with slower population growth rates.

Rapid population growth is not simply something that agriculture can respond to by producing extra food. It also has a

Table 5. FUTURE POPULATION RATIOS, LOW POPULATION PROJECTION

YEAR	INPUT LEVEL	AFRICA	SW ASIA	SE ASIA	SOUTH AMERICA	CENTRAL AMERICA
2000	Low	0.6	0.5	0.95	0.5	0.6
	Intermediate	2.2	0.65	1.7	2.0	1.1
	High	6.2	0.9	2.4	4.8	2.5
2025	Low	0.6	0.6	1.5	0.55	0.7
	Intermediate	1.8	0.6	1.7	2.0	1.0
	High	5.2	0.75	2.4	4.6	2.1

The figures show the fraction of expected populations that regional land resources could feed with given input level.

powerful impact on agriculture, and can often stand in the way of efforts to improve the productivity of the land. Throughout Asia, Latin America and those parts of Africa where individual private landownership prevails, smallholdings are being subdivided by inheritance — and the more (male) children the owner has, the more his holding will be subdivided on his death. In many areas, a large proportion of holdings are now too small to support a family, let alone to generate any surplus that could be invested in more inputs such as fertilizer, land improvements, or small-scale machinery. Rapid population growth also adds copiously to the rural labour supply in many areas, and may, in the absence of investment, create more jobseekers in rural areas, depress wages and slow down the rate of mechanization of agriculture.

Perhaps even more serious is the effect of over-rapid population growth on ecologically sensitive areas. The increasing density of population has forced farmers in poor areas to cut back on fallow periods, essential to restore soil fertility, without applying extra fertilizer to compensate. As a result, soil fertility gradually deteriorates and yields may fall, forcing farmers to cut fallow periods back even further.

The fuelwood crisis also has a powerful impact on the environment. The number of people without adequate fuelwood supply, or who are forced to overcut the stock of trees to get enough, is expected to rise from 1 395 million in 1980 to 2 986 million by the end of the century. Four fifths of that rise will be due to population increase. People facing shortages are forced to forage over a wider and wider area, cutting from and sometimes destroying the scattered stands of trees by roads and rivers, on slopes and in drier areas, where they are essential to soil conservation. Once these stands are removed, soil erosion and water run-off increase upstream, while downstream flooding and silting up of irrigation systems may result. In semi-arid areas, removal of tree cover can contribute to desertification (Ref. 16).

The pressure of population in farming areas (often intensified by gross inequities in landholdings) forces many poor people to begin cultivating more marginal areas, such as excessive slopes, semi-arid areas, or areas with unreliable rainfall. Without proper conservation measures, this can result in erosion and desertification, and also exposes migrants to the threat of drought and recurrent famine.

The bulk of migration in developing countries is, of course, toward urban areas. Rapid population growth in rural areas fuels the exodus to the cities, although there are other causes such as higher city incomes and the relative neglect by governments of adequate levels of investment in rural areas. Rural migrants to the cities are often younger and better educated than those they leave behind, so the productivity and openness to technological change of farmers may be diminished.

Reducing the rates of population growth can help, along with other measures, to slow down or halt these processes. It can help to reduce labour surpluses where these exist, and act as an incentive to mechanization. It can reduce the pressure on the environment. For countries that are not self-sufficient in food, it can reduce the amount of food imports required.

In a number of cases, slower population growth would make

all the difference between a country being critical or non-critical with a given level of inputs. The main study's results are based on the United Nations' *medium* population projection, which represents the most probable outcome on present trends. The UN also publishes a low projection which could come about if fertility rates declined faster than expected (Ref. 30). Six countries — five of them in Africa — would be removed from the critical list for low inputs in 2000 if they could reduce their population growth rate in line with the low projection rather than the medium. These are Zimbabwe, Benin, Mali, Swaziland, Sierra Leone and the Dominican Republic. Another three countries — Mauritania, Bangladesh and the Syrian Arab Republic — would cease to be critical at the intermediate-input level.

However, all the critical countries would find the task of feeding their populations — either from their own lands, or from imports — lightened if they achieved lower population growth. On the medium projection, for example, with low inputs Africa could feed only 41 percent of its year 2025 population, but it could feed 58 percent if the low projection were realized. In the latter case, the continent would need to raise its food production by 72 percent above the low-input level to meet its food requirements — but it would have to raise production by 144 percent to meet the needs of the *medium* projection population. If it were to meet the food deficit by imports, with the lower population it would have to import 42 percent of its needs, but with the medium projection it would need to import 59 percent.

After adjustment for cultivated land and non-food crops, Central America will be critical in 2000 at the intermediate-input level on the *medium* population projection. However, if it achieved reduced population growth in line with the *low* projection, it would no longer be critical, and instead of having to import 6 percent of its requirements, it would have the potential to *export* 6 percent.

Similar points apply to all regions and all countries. In cases of deficit, the imports or the increases in agricultural effort required to meet food needs are lower for the low population projection. For surplus regions, the potential surplus for export is

FIGURE 13 POPULATION: LOW AND HIGH PROJECTIONS (millions)

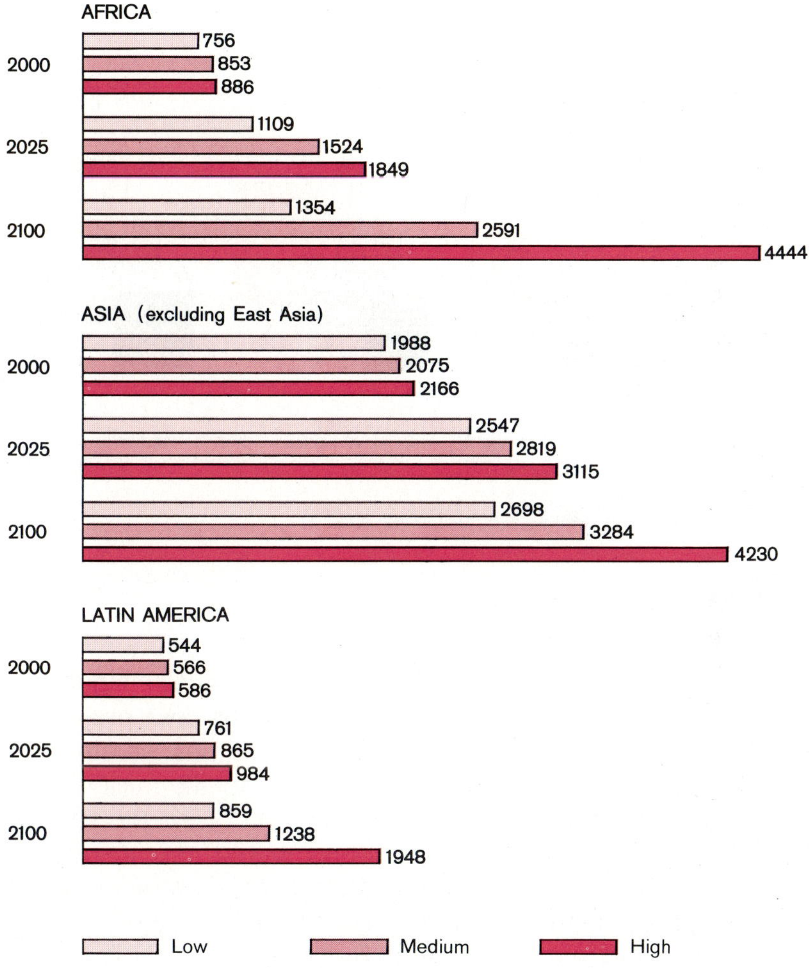

SOURCE: WORLD POPULATION PROSPECTS AS ASSESSED IN 1980, UN POPULATION DIVISION, 1982

greater. The contribution of reduced population growth to increased food security is clear.

Figure 13 indicates the population ratios that would apply if the UN low projection became reality.

FIGURE 14

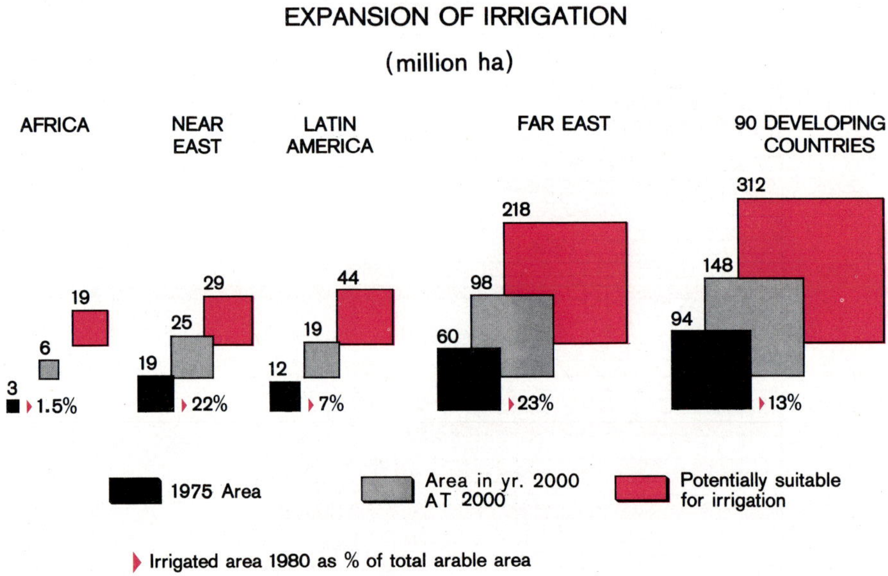

SOURCE: AT 2000 DATA

The *Agriculture: Toward 2000* study found equally significant benefits from reduced population growth. On the *low* projection, for example, the cereal imports of the 90 countries studied would fall from 48 million tonnes in 1979-81 to 43 million tonnes in 2000 (on the optimistic Scenario A). With the *medium* population projection, by contrast, imports would *rise* to 53 million tonnes. On the *medium* projection, Africa would remain a net importer of agricultural produce; on the *low* projection, it could become a net exporter again. Reduced population growth would also help to cut the incidence of malnutrition, from an expected 685 million by the end of the century, on present trends, to 647 million — 38 million fewer (Ref. 34).

The benefits of lowered population growth increase with time — just as do the costs of rapid growth rates. Over the long term, massive differences build up (see Fig. 13). On the low projection, for example, the population of Africa in 2000 would be 756 million, or 100 million less than the medium projection. By 2025, the low projection would be 1 109 million, against 1 542 million on the medium projection. The difference of 433 million is more than the entire 1975 population of the continent. The differences for individual countries are of similar proportions. Kenya (present population: 18.6 million) would have no less than 82 million inhabitants by 2025 on the medium projection, but only 53 million on the low — roughly equal to the numbers its entire land resources could support with high inputs, optimum crop mixes, and all land under food crops. Burundi will have 9.3 million people in 2025 on the low projection — again, this is about the number its entire land could support using high inputs. On the medium projection, however, Burundi's population in 2025 would be 13.3 million (Ref. 20).

Because of the long-term build-up effects of small differences in present population growth rates, it is important to act as soon as possible to interrupt the momentum of rapid growth. For most developing countries, attaining the UN low projection should not be too difficult. The birth rate for year 2000 for most regions would need to be between 1.5 and 2.9 births per 1 000 population, lower than in the medium projection, and would involve a drop of about 10 births per 1 000 over 20 years. For Africa, the task would be more challenging, involving a drop from 45.6 per 1 000 in 1980 to 30.3 per 1 000 in 2000 — 15 points in 20 years. This is less than has been achieved, for example, by Costa Rica, Cuba, China or the Republic of Korea. However, even if the full reduction proves impossible, any reduction at all will prove beneficial, especially in the long run.

As with agriculture, there is a wide range of measures governments can take toward reducing their birth rates. Those countries that have achieved the most rapid declines have generally pursued strategies that placed great emphasis on the development of human resources, through rapid and widespread improvements in health care and education. Many have also followed strategies of growth with redistribution, in which the incomes of the poorest strata have risen somewhat faster than average. Improvements in the status of women, and in

FIGURE 15

CRITICAL COUNTRIES' ATTITUDES TO FERTILITY
Risk and percentage of year 2000 population that can be supported with low inputs

Government view on fertility, 1980 rate	Higher risk[1]		Lower risk[2]	
	0-50 percent	50-100 percent	0-50 percent	50-100 percent
Too low: measures implemented to raise fertility	IRAQ		ISRAEL[3]	
Satisfactory: measures implemented to maintain	NIGER MAURITANIA	UPPER VOLTA BENIN TOGO MALI	KUWAIT OMAN QATAR SAUDI ARABIA, KINGDOM OF UNITED ARAB EMIRATES SINGAPORE[3]	
Satisfactory: no government intervention	BURUNDI[4] ETHIOPIA SOMALIA DEM. YEMEN YEMEN NIGERIA	BHUTAN MALAWI SYRIA	CAPE VERDE[3] LEBANON[3]	
Too high: no government intervention appropriate	ALGERIA COMOROS RWANDA JORDAN AFGHANISTAN	SENEGAL SIERRA LEONE GUATEMALA[4]	BAHRAIN SAINT LUCIA[3]	BAHAMAS[3] DOMINICAN REP.[3] ZIMBABWE
Too high: measures implemented to reduce	KENYA LESOTHO[4] UGANDA EL SALVADOR HAITI[4]	BOTSWANA MOROCCO BANGLADESH SWAZILAND	BARBADOS[3] MAURITIUS[3]	IRAN SRI LANKA[3] VIET NAM[4] PHILIPPINES[3] JAMAICA[3] TRINIDAD AND TOBAGO[3] TUNISIA[4]

[1] Scoring 6 or more on exposure factors
[2] Scoring 5 or less. — [3] Fertility rates below 5. — [4] Fertility rates below 6.

education, health care and expansion of formal work opportunities for women also have a strong impact in reducing the population fertility rate. All these measures, of course, are well worth pursuing for their own sake, regardless of the effect on fertility. They can also be pursued using low-cost methods, with appropriate technology, community involvement, and the use of auxiliary personnel, long before economic development makes more sophisticated approaches possible.

Of course, improvements in health care reduce mortality and infant mortality and, as population fertility may take longer to adjust downward, there may be a temporary increase in the rate of population growth. Hence, approaches are needed which reduce the lag in fertility decline to a minimum. The encouragement of breast-feeding, for example, or of wider child spacing improves child health and reduces fertility at the same time.

The widespread availability of effective methods of family planning, through health personnel, community distributors, local shops, etc., is important — and valuable in itself, allowing families to decide on the number of children they wish to have. Government support for family planning helps to create a climate where limitation of family size is seen as desirable.

Yet, government support is often lacking where it is most urgently needed. Figure 15 shows the results of the United Nations survey of government attitudes to population fertility in 1980. Replies were available from 56 countries that will be critical in 2000 with low inputs (Ref. 35). Among countries that would be able to support *less* than half their expected populations, only 44 percent of governments thought their country's fertility rate was too high — against 71 percent in the case of countries that could support *more* than half their populations. Only 52 percent of the *higher risk* countries (with six or more risk factors affecting their food situation) — see Fig. 8 — thought that their national fertility rate was too high, compared with 61 percent among those with fewer risk factors.

Indeed, there were 16 higher risk countries, critical with low inputs in year 2000, which considered their fertility rates satisfactory or too low. Twelve of these had no significant fuel mineral reserves or manufactured exports. Nine of the latter were in Africa: Benin, Burundi, Ethiopia, Malawi, Mali, Mauritania, Somalia, Togo and the Upper Volta. The other three were Democratic Yemen, Yemen (although, here, migrant labour to neighbouring oil states does provide significant foreign exchange) and Bhutan.

However, in many of the countries whose governments do consider their birth rates too high, family-planning programmes are often ineffective and have had little or no impact so far on birth rates.

It is to be hoped that the results of the land resources study will underline, for those higher risk countries which do not have secure sources of foreign exchange, the crucial importance of effective population policies and programmes, as well as of agricultural development, in balancing the food-population equation over the coming decades.

Conclusion

The results of the Land Resources for Future Populations Project are, because of the limitations of data and methodology, a first approximation only. In future, as FAO's 1983 Conference recommended, the methodology must be refined to include a wider range of crops, the detailed potential of forests, fisheries and irrigation, and national and international economic issues such as trade, aid, access to technology and the range of equity and participation issues raised by the World Conference on Agrarian Reform and Rural Development. Much more detailed national-level studies are required, and some are already under way. These can provide a more reliable basis for national-level planning and coordination of land use, agricultural development, population planning, and other aspects of development.

There are no easy solutions and no global panaceas. Each country will have to devise its own package of measures, never forgetting the central importance of conservation of the environment, the development of human resources, equity and people's participation in development programmes. Without attention to these factors, no plan or purely technical solution stands much chance of success.

One requirement, however, does stand out clearly: the need for contingency planning over a much longer time-span than has been usual so far. Since its inception, economic planning by governments has rarely looked beyond the next five years.

At this point in human history, such a time-span is utterly inadequate. The human race is now at a crucial turning-point. The next 40 years will see the largest absolute increase in human numbers of all time. There will be a gradual depletion of oil reserves and the need for transition to renewable forms of energy. The introduction of new technologies of unprecedented productivity will set the problem of distributing employment and the fruits of production equitably. The human race will either suffer retribution for its failure to control land degradation and the arms race — or be rewarded for its success in halting these destructive processes.

The long-term outcome of these challenges is, of course, uncertain — because it depends in large part on present actions. For survival as a species with any semblance of general welfare, we must begin to plan our responses to these challenges immediately. Clearly, the longer we delay preparations, the more likely it is that the more unpleasant scenarios will come to pass.

The full support of the international community should be provided, for the welfare of people in individual countries is now a matter of global concern. Working together to improve the long-term balance of populations and the possibilities of food provision will undoubtedly help to contribute to world peace.

References

1. *FAO-Unesco Soil Map of the World* (1:5 000 000), 10 vols. Paris, Unesco. 1971-81.
2. *Agro-ecological Zones Project*, Vols 1-4. World Soil Resources Report 48/1-4. Rome, FAO. 1978-80.
3. *Potential population-supporting capacities of lands in the developing world*. FPA/INT/513. Rome, FAO. 1982.
4. *Agro-ecological Zones Project. Op. cit.*, Vol. 1.
5. *Report on the Second FAO/UNFPA Expert Consultation on Land Resources for Populations of the Future*. Rome, FAO. 1980.
6. *Agriculture: Toward 2000*. Rome, FAO. 1981.
7. *FAO Fertilizer Yearbook 1981*. Rome, FAO.
8. *Provisional Statistics on Fertilizer Use by Major Crops*. ESS/MISC/1980/3. Rome, FAO. 1980. (Mimeo)
9. See Ref. 5, p. 148-149.
10. *FAO Production Yearbook 1981*. Rome, FAO.
11. Calculated from *World Oil*, 15 August 1983.
12. See, for example, A.R. Khan *et al. Poverty and landlessness in rural Asia*. ILO. 1977.
13. *Agriculture: Toward 2000*, first version, C/79/24, p. 64. Rome, FAO. 1979.
14. *The State of Food and Agriculture 1980*, p. 18. Rome, FAO. 1981.
15. *Map of the Fuelwood Situation in Developing Countries*. Rome, FAO, 1981.
16. *Fuelwood supplies in the developing countries*. FAO Forestry Paper 42. Rome, FAO. 1982.
17. *Food Balance Sheets 1975-77*. Rome, FAO. 1983.
18. *Fourth World Food Survey*, p. 53, 62-63. Rome, FAO. 1977.
19. *The State of Food and Agriculture 1981*. Rome, FAO. 1982.
20. *World population prospects as assessed in 1980*. Department of Economic and Social Affairs, UN, New York. 1981.

21. *Agriculture: Toward 2000, Statistical Compendium*. Rome, FAO. 1983. (Mimeo)
22. Calculated from Ref. 10.
23. Calculated from Ref. 21.
24. *Explanatory Note and UN Desertification Map of the World*. UN Desertification Conference, Nairobi, Kenya. A/Conf 74/2. 1977.
25. Cited in Ref. 13.
26. Soil and Water Conservation. COAG/81/8. Rome, FAO. 1980.
27. UN Economic and Social Council, Committee on Natural Resources. E/C/1983/11. 1983.
28. *Evaluation of the irrigation potential for the African continent*. R.G. Thomas and K. Siegert. Rome, FAO. 1981. (Unpublished)
29. *Global 2000, Report to the President*. Technical Report. Washington, DC. 1980.
30. *Understanding agriculture: growth, structure and challenge*. Background paper for *AT 2000*. G. Parthasarathy. Rome, FAO. 1979. (Mimeo)
31. *Agrarian structure and farm productivity in developing countries*. R.A. Berry and W.R. Cline. Johns Hopkins. 1979.
32. See Ref. 5, p. 6-9.
33. *Postharvest food losses in developing countries*. National Academy of Sciences, Washington, DC. 1978.
34. *Agriculture from the perspective of population growth*. Alexandratos, Bruinsma, Yotopoulos. FAO Economic and Social Development Paper 30. Rome, FAO. 1983.
35. *World population trends and policies - 1981 Monitoring Report*, p. 107-110. Population Studies No. 70, Department of International Economic and Social Affairs, UN, New York. ST/ESA/SER A/79. 1982.

Appendix 1

Present and projected populations and ratios to present and projected population-supporting capacities by location

Africa, Southwest Asia, South America, Central America and Southeast Asia

AFRICA

Location	Land[1] area (mill. ha)	Year[2] 1975 population (mill.)	Year 1975 population density	..... Year 1975 ratios ..... Low inputs	Intermed. inputs	High inputs	Year[3] 2000 population (mill.)	Year 2000 population density	..... Year 2000 ratios ..... Low inputs	Intermed. inputs	High inputs
Cape Verde	0.4	0.3	0.74	*<0.01	*<0.01	*0.02	0.4	1.06	*<0.01	*<0.01	*<0.01
Western Sahara	26.6	0.1	<0.01	*<0.01	*<0.01	*<0.01	0.2	<0.01	*<0.01	*<0.01	*<0.01
Rwanda	2.5	4.1	1.62	*0.18	*0.88	1.99	9.3	3.53	*0.08	*0.38	*0.85
Totals for Class 1	29.6	4.5	0.15	*0.16	*0.80	1.81	9.7	0.33	*0.88	*0.36	*0.79
Burundi	2.6	3.9	1.52	*0.22	1.17	2.66	7.8	3.03	*0.10	*0.54	1.23
Niger	126.1	4.6	0.04	*0.17	*0.56	9.43	9.7	0.08	*0.14	*0.32	4.48
Réunion	0.3	0.5	1.92	*0.20	*1.00	3.33	0.7	2.73	*0.14	*0.67	2.23
Kenya	57.0	13.5	0.24	*0.27	*0.90	3.90	37.6	0.59	*0.17	*0.40	51.7
Lesotho	3.0	1.2	0.39	*0.46	1.15	2.64	2.1	0.69	*0.26	*0.65	1.48
Comoros	0.2	0.3	1.40	*0.42	1.78	3.48	0.6	2.07	*0.27	1.16	2.26
Algeria	237.8	15.7	0.07	*0.45	1.57	2.99	36.0	0.15	*0.33	*0.79	1.37
Somalia	63.6	3.2	0.05	*0.40	*0.75	2.18	7.2	0.10	*0.34	*0.51	1.23
Ethiopia	120.8	28.8	0.24	*0.59	2.38	10.73	55.3	0.46	*0.36	1.28	5.55
Nigeria	91.2	65.7	0.72	*0.82	3.32	11.27	150.0	1.63	*0.37	1.39	4.71
Namibia	82.3	0.9	<0.01	*0.84	2.40	36.56	1.7	0.02	*0.43	1.24	18.82
Uganda	20.0	11.3	0.57	*0.97	3.88	13.60	25.4	1.23	*0.45	1.75	6.08
Mauritania	103.1	1.4	<0.01	*0.35	1.42	5.92	2.9	0.03	*0.46	*0.97	3.08
Mauritius	0.2	0.9	4.85	*0.71	1.27	1.01	1.3	7.09	*0.47	*0.86	*0.79
Totals for Class 2	908.1	151.8	0.17	*0.65	2.53	9.29	331.4	0.36	*0.34	1.15	4.12
Upper Volta	27.1	6.1	0.22	*0.90	4.36	22.48	11.8	0.44	*0.51	2.26	11.45
Malawi	8.8	5.2	0.59	1.27	4.63	10.88	12.0	1.35	*0.61	2.01	4.62
Botswana	60.0	0.7	<0.01	1.27	6.67	64.61	1.4	0.02	*0.63	3.32	32.13
Tunisia	16.3	5.6	0.34	*0.45	1.12	2.14	9.6	0.59	*0.70	1.06	1.62
Morocco	44.4	17.3	0.39	*0.73	1.55	2.69	36.1	0.81	*0.78	1.15	1.65
Senegal	19.6	5.0	0.25	*0.89	3.89	21.00	9.7	0.49	*0.79	2.31	10.93
Togo	5.6	2.3	0.42	1.69	7.92	22.47	5.0	0.90	*0.82	3.65	10.16
Sierra Leone	7.2	3.0	0.42	1.69	9.14	16.43	6.1	0.84	*0.89	4.52	8.12
Swaziland	1.7	0.5	0.28	1.75	7.40	13.51	0.9	0.54	*0.89	3.77	6.88
Mali	124.0	5.8	0.05	1.34	6.04	29.44	11.6	0.09	*0.90	3.21	14.73

Benin	11.3	3.0	0.27	2.08	9.34	33.60	6.5	0.58	*0.95	4.28	15.40
Zimbabwe	38.6	6.2	0.16	1.65	7.80	29.88	14.7	0.36	*0.95	3.64	13.21
Totals for Class 3	364.5	60.9	0.17	1.10	4.44	15.96	124.7	0.34	*0.78	2.36	7.86
Ghana	22.6	10.0	0.44	2.04	8.86	24.40	21.2	0.94	1.03	4.19	11.25
Tanzania	88.9	15.4	0.17	2.29	9.37	32.55	34.0	0.38	1.12	4.30	14.72
Libya	175.9	2.4	<0.01	*0.74	1.51	2.71	5.8	0.03	1.20	1.47	1.89
Gambia	1.1	0.5	0.46	1.49	6.44	21.47	1.0	0.90	1.21	3.63	11.10
Egypt	99.1	36.9	0.37	1.67	1.67	1.67	64.7	0.65	1.74	1.74	1.74
Chad	127.5	4.3	0.03	3.15	16.10	70.80	7.1	0.06	1.78	8.77	38.30
São Tomé and Principe	0.1	0.1	0.87	1.98	7.21	9.71	0.1	0.96	1.79	6.52	8.79
Guinea	24.6	4.4	0.18	3.39	12.22	39.54	8.2	0.33	1.82	6.54	21.16
Totals for Class 4	539.8	74.1	0.14	2.00	5.75	17.64	142.8	0.26	1.47	3.39	9.47
Mozambique	78.3	9.2	0.12	4.14	17.01	58.07	18.7	0.23	2.17	8.65	29.33
Sudan	250.5	16.0	0.06	3.71	14.91	64.72	31.3	0.12	2.58	8.30	33.73
Guinea-Bissau	3.6	0.5	0.15	4.40	17.23	55.97	0.8	0.23	2.72	10.65	34.62
Liberia	11.1	1.6	0.14	6.03	30.35	80.52	3.5	0.31	2.77	13.62	36.14
Ivory Coast	31.5	6.7	0.21	7.15	24.56	55.57	13.9	0.44	3.41	11.65	26.98
Madagascar	58.0	7.7	0.13	6.29	25.03	72.88	15.2	0.26	3.67	12.97	37.03
Angola	123.9	6.2	0.05	8.57	45.01	150.14	12.4	0.09	4.51	23.52	78.98
Zambia	74.3	4.8	0.06	10.17	44.88	159.38	11.3	0.14	4.68	20.63	73.31
Totals for Class 5	631.2	52.7	0.08	5.84	24.74	82.65	105.0	0.17	3.21	12.60	41.55
Cameroon, Rep. of	47.2	7.5	0.16	10.25	27.93	81.81	13.0	0.28	5.89	16.03	46.92
Zaire	232.2	24.5	0.11	11.91	52.27	117.83	46.2	0.20	6.29	27.77	62.26
Eq. Guinea	2.8	0.3	0.12	12.08	35.65	104.44	0.6	0.20	6.93	20.44	59.87
Totals for Class 6	282.2	32.3	0.11	11.53	46.46	109.35	59.7	0.21	6.21	25.15	58.90
Centr. African Rep.	62.3	2.0	0.03	22.59	106.68	301.12	3.9	0.06	12.45	58.97	105.94
Congo	34.2	1.4	0.04	29.90	120.11	299.88	2.4	0.07	16.44	65.76	164.02
Gabon	26.2	0.5	0.02	80.19	243.44	538.40	0.8	0.03	55.53	168.58	372.84
Totals for Class 7	122.7	3.9	0.03	32.93	129.86	332.73	6.8	0.06	18.65	73.52	188.07
Africa totals	2 878.1	380.2	0.13	2.95	11.55	33.85	780.1	0.27	1.61	5.75	16.50

[1] Land areas derived from FAO/Unesco Soil Map and exclude areas mapped as bodies of water. - [2] UN data for 1975, millions of persons, UN. 1979a. - [3] Projected UN data for 2000, millions of persons. Medium variant. UN. 1979a. - * Asterisk indicates potential population-supporting capacity less than present and/or projected population. Original data to three decimal places. Apparent discrepancies in some totals due to rounding off. This table has been sorted according to increasing ratios, by country, related to low-input year 2000 ratios.

SOUTHWEST ASIA

Location	Land[1] area (mill. ha)	Year[2] 1975 population (mill.)	Year 1975 population density	Year 1975 ratios Low inputs	Year 1975 ratios Intermed. inputs	Year 1975 ratios High inputs	Year[3] 2000 population (mill.)	Year 2000 population density	Year 2000 ratios Low inputs	Year 2000 ratios Intermed. inputs	Year 2000 ratios High inputs
Bahrain	0.1	0.3	4.27	<*0.01	<*0.01	<*0.01	0.5	8.97	<*0.01	<*0.01	<*0.01
Qatar	1.1	0.2	0.15	<*0.01	<0.01	<*0.01	0.4	0.39	<*0.01	<*0.01	<*0.01
Oman	21.3	0.8	0.04	<*0.01	<*0.01	<*0.01	1.6	0.08	<*0.01	<*0.01	<*0.01
Kuwait	1.6	1.0	0.63	<*0.01	*0.01	*0.01	3.2	1.98	<*0.01	<*0.01	<*0.01
United Arab Emirates	8.4	0.6	0.07	*0.18	*0.18	*0.18	1.6	0.20	*0.08	*0.08	*0.08
Totals for Class 1	33.5	2.8	0.08	*0.04	*0.04	*0.04	7.4	0.22	*0.02	*0.02	*0.02
Jordan	9.8	2.7	0.28	*0.24	*0.42	*0.52	5.9	0.60	*0.16	*0.23	*0.27
Saudi Arabia	215.0	7.2	0.03	*0.21	*0.27	*0.67	15.6	0.07	*0.17	*0.19	*0.38
Lebanon	1.0	2.8	2.69	*0.25	*0.42	*0.49	4.9	4.70	*0.23	*0.30	*0.33
Democratic Yemen	28.2	1.6	0.06	*0.58	*0.71	1.17	3.4	0.12	*0.30	*0.37	*0.58
Israel	2.1	3.4	1.67	*0.38	*0.62	*0.89	5.6	2.72	*0.31	*0.41	*0.50
Yemen Arab Rep.	19.5	5.3	0.27	*0.48	*0.69	1.45	10.0	0.51	*0.35	*0.47	*0.91
Iraq	43.5	11.0	0.25	*0.82	1.78	3.17	24.3	0.56	*0.38	*0.74	1.23
Afghanistan	64.8	19.3	0.30	*0.62	*0.77	1.06	36.6	0.57	*0.43	*0.51	*0.66
Totals for Class 2	383.7	53.4	0.14	*0.54	*0.86	1.42	106.2	0.28	*0.34	*0.48	*0.72
Syria	18.5	7.3	0.40	*0.72	1.75	2.66	16.3	0.88	*0.61	*0.96	1.33
Iran	164.8	32.7	0.20	*0.86	1.16	1.65	65.4	0.40	*0.63	*0.78	1.02
Totals for Class 3	183.3	40.1	0.22	*0.84	1.27	1.84	81.7	0.45	*0.63	*0.81	1.08
Turkey	76.9	40.1	0.52	1.12	1.91	2.96	69.4	0.90	1.34	1.75	2.29
Totals for Class 4	76.9	40.1	0.52	1.12	1.91	2.96	69.4	0.90	1.34	1.75	2.30
Southwest Asia totals	677.4	136.3	0.20	*0.79	1.27	1.96	264.7	0.39	*0.68	*0.90	1.23

SOUTH AMERICA

Chile	75.7	10.2	0.13	1.28	2.50	4.38	14.9	0.20	1.75	2.58	3.83
Totals for Class 4	75.7	10.2	0.13	1.28	2.51	4.39	14.9	0.20	1.75	2.57	3.82
Ecuador	28.4	6.9	0.24	3.76	16.20	25.54	14.6	0.51	2.39	8.20	12.47
Brazil	851.3	109.7	0.13	5.45	27.06	65.02	212.5	0.25	2.93	14.01	33.44
Peru	128.5	15.5	0.12	5.06	12.32	39.00	29.5	0.23	3.10	6.91	20.98
Colombia	103.9	23.8	0.23	5.30	16.58	37.59	42.5	0.41	3.27	9.57	21.22
Venezuela	88.2	12.6	0.14	6.22	19.01	58.74	25.7	0.29	3.45	9.67	29.03
Argentina	278.9	25.4	0.09	5.72	16.75	36.41	32.8	0.12	4.95	13.28	28.49
Totals for Class 5	1 479.2	194.0	0.13	5.43	22.35	54.06	357.5	0.24	3.19	12.30	29.37
Uruguay	17.8	2.8	0.16	6.51	19.17	47.10	3.4	0.19	5.47	15.87	38.80
Paraguay	40.7	2.6	0.06	12.69	44.43	117.03	5.3	0.13	6.51	22.08	58.11
Bolivia	109.9	4.9	0.04	15.58	61.22	146.93	9.3	0.08	8.45	32.58	77.86
Totals for Class 6	168.3	10.4	0.06	12.44	45.73	112.71	18.0	0.11	7.33	26.37	64.76
Guyana	21.5	0.8	0.04	33.84	132.57	324.11	1.3	0.06	22.33	84.65	206.48
Suriname	16.3	0.4	0.02	76.00	362.04	807.54	0.7	0.04	39.72	185.79	413.26
French Guiana	9.1	0.1	0.01	388.29	999.00	999.00	0.1	0.01	209.00	753.38	826.77
Totals for Class 7	46.9	1.2	0.03	65.25	247.24	505.44	2.1	0.04	38.68	156.32	310.32
South America totals	1 770.2	215.8	0.12	5.91	23.92	57.22	392.6	0.22	3.51	13.34	31.51

[1] Land areas derived from FAO/Unesco Soil Map and exclude areas mapped as bodies of water. - [2] UN data for 1975, millions of persons, UN. 1979. - [3] Projected UN data for 2000, millions of persons. Medium variant. UN. 1979. - * Asterisk indicates potential population-supporting capacity less than present and/or projected population. Original data to three decimal places. Apparent discrepancies in some totals due to rounding off. This table has been sorted according to increasing ratios, by country, related to low input year 2000 ratios.

CENTRAL AMERICA

Location	Land[1] area (mill. ha)	Year[2] 1975 population (mill.)	Year 1975 population density	Year 1975 ratios: Low inputs	Year 1975 ratios: Intermed. inputs	Year 1975 ratios: High inputs	Year[3] 2000 population (mill.)	Year 2000 population density	Year 2000 ratios: Low inputs	Year 2000 ratios: Intermed. inputs	Year 2000 ratios: High inputs
Neth. Antilles	0.1	0.2	2.51	*<0.01	*<0.01	*<0.01	0.4	4.05	*<0.01	*<0.01	*<0.01
Barbados	0.0	0.2	5.70	*0.09	0.52	1.20	0.3	6.91	*0.06	*0.39	*0.90
Totals for Class 1	0.1	0.5	3.43	*0.04	*0.25	*0.57	0.7	4.93	*0.03	*0.16	*0.36
Martinique	0.1	0.3	2.95	*0.26	1.07	2.76	0.4	3.26	*0.23	*0.95	2.45
Guadeloupe	0.2	0.3	1.83	*0.39	2.98	5.10	0.4	2.14	*0.33	2.50	4.28
Antigua and Barbuda	0.0	0.1	1.66	*0.40	*0.92	3.09	0.1	1.93	*0.34	*0.78	2.62
El Salvador	2.1	4.1	1.97	*0.49	1.21	3.81	8.7	4.13	*0.35	*0.64	1.70
Haiti	2.8	5.2	1.86	*0.51	1.20	2.23	9.9	3.56	*0.40	*0.72	1.16
Windward Is.	0.2	0.4	1.81	*0.58	2.06	6.11	0.4	2.09	*0.49	1.75	5.21
Totals for Class 2	5.4	10.4	1.92	*0.49	1.28	3.12	19.8	3.67	*0.38	*0.75	1.57
Trinidad and Tobago	0.5	1.1	2.11	*0.50	2.49	4.15	1.4	2.68	*0.57	2.06	3.26
Puerto Rico	0.9	3.1	3.51	*0.39	*0.97	2.00	4.4	4.97	*0.58	*0.93	1.51
Jamaica	1.1	2.0	1.86	*0.77	2.08	4.54	2.9	2.62	*0.76	1.61	3.21
Bahamas	1.0	0.2	0.20	1.29	15.73	58.91	0.3	0.33	*0.79	9.62	36.02
Guatemala	10.8	6.2	0.58	1.25	3.58	12.26	12.7	1.18	*0.89	1.98	6.05
Dominican Rep.	4.8	5.2	1.08	1.14	2.56	5.28	9.3	1.93	*0.95	1.70	3.09
Totals for Class 3	19.1	17.9	0.94	*0.97	2.73	7.61	31.1	1.62	*0.84	1.80	4.45
Mexico	197.3	59.2	0.30	1.69	4.08	10.71	132.3	0.67	1.38	2.40	5.26
Honduras	11.2	3.1	0.28	2.97	7.41	23.86	7.0	0.62	1.48	3.43	10.58
Totals for Class 4	208.5	62.3	0.30	1.75	4.25	11.35	139.3	0.67	1.38	2.45	5.53
Cuba	11.4	9.3	0.82	1.20	2.23	8.57	12.7	1.11	2.22	2.85	6.67
Costa Rica	5.1	2.0	0.39	3.05	9.65	26.26	3.4	0.67	2.32	6.06	15.27
Panama	7.6	1.7	0.22	5.24	17.71	44.30	2.8	0.37	3.30	10.54	26.21
Nicaragua	12.1	2.3	0.19	6.38	18.33	51.33	5.2	0.43	3.30	8.55	22.89
Totals for Class 5	36.2	15.3	0.42	2.67	7.33	21.27	24.1	0.66	2.60	5.43	13.66
Belize	2.3	0.1	0.06	15.28	69.20	175.05	0.2	0.10	9.03	40.89	103.45
Totals for Class 6	2.3	0.1	0.06	15.18	68.74	173.90	0.2	0.10	9.22	41.75	105.62
Central America totals	271.6	106.5	0.39	1.64	4.21	11.51	215.2	0.79	1.35	2.57	6.01

SOUTHEAST ASIA

Singapore	0.1	2.3	38.79	*<0.01	*<0.01	*<0.01	3.1	53.36	*<0.01	*<0.01	*<0.01
Totals for Class 1	0.1	2.3	37.50	*<0.01	*<0.01	*<0.01	3.1	51.50	*<0.01	*<0.01	*<0.01
Viet Nam	33.0	46.5	1.41	*0.90	2.67	4.90	79.3	2.41	*0.74	1.69	2.92
Bhutan	4.7	1.2	0.25	1.34	3.83	6.92	2.0	0.43	*0.76	2.16	3.91
Sri Lanka	6.6	13.6	2.07	*0.83	2.42	3.55	20.1	3.06	*0.77	1.71	2.35
Bangladesh	14.4	76.6	5.32	*0.46	1.31	2.10	153.3	10.65	*0.79	*0.97	1.21
Philippines	30.0	43.8	1.46	1.08	3.80	6.10	83.4	2.78	*0.87	2.14	3.14
Totals for Class 3	88.6	181.7	2.05	*0.76	2.36	3.92	338.2	3.82	*0.80	1.48	2.17
India	320.4	618.7	1.93	*0.81	1.83	3.54	1 036.7	3.23	1.00	1.74	2.53
Nepal	14.1	12.7	0.90	*0.94	2.65	4.06	22.4	1.59	1.05	1.77	2.46
Burma	67.6	31.2	0.46	1.82	5.78	11.52	55.1	0.81	1.19	3.35	6.44
Thailand	51.4	41.4	0.80	1.44	4.00	8.86	76.1	1.48	1.38	2.65	5.05
Democratic Kampuchea	17.6	8.1	0.46	2.39	9.35	19.35	13.4	0.76	1.52	5.66	11.58
Laos	23.7	3.3	0.14	2.54	8.41	22.19	5.7	0.24	1.58	4.91	12.79
Indonesia	191.9	135.9	0.71	2.10	7.25	10.60	221.0	1.15	1.58	4.65	6.65
Pakistan	88.7	70.3	0.79	1.14	1.24	1.37	145.0	1.63	1.59	1.63	1.68
Brunei	0.6	0.2	0.28	3.26	28.85	34.60	0.3	0.53	1.71	15.11	18.16
Totals for Class 4	776.1	921.8	1.19	1.11	2.92	5.14	1 575.6	2.03	1.17	2.28	3.40
Malaysia	32.8	11.9	0.36	7.30	21.50	19.75	20.1	0.61	4.56	12.91	11.79
Totals for Class 5	32.8	11.9	0.36	7.31	21.52	19.76	20.1	0.61	4.56	12.91	11.79
Southeast Asia totals	897.6	1 117.7	1.25	1.12	3.02	5.09	1 937.1	2.16	1.14	2.25	3.27
Five Regions Total	6 494.9	1 956.6	0.30	2.03	6.94	16.56	3 589.7	0.55	1.56	4.16	9.25

[1] Land areas derived from FAO/Unesco Soil Map and exclude areas mapped as bodies of water. - [2] UN data for 1975, millions of persons, UN. 1979. - [3] Projected UN data for 2000, millions of persons. Medium variant. UN. 1979. - * Asterisk indicates potential population-supporting capacity less than present and/or projected population. Original data to three decimal places. Apparent discrepancies in some totals due to rounding off. This table has been sorted according to increasing ratios, by country, related to low input year 2000 ratios.

Appendix 2

Statistical analyses of the 64 countries and territories critical for low inputs in the year 2000

1. Low inputs: risk factor score
2. Summary of risk factors
3. The most critical countries, year 2000

1. LOW INPUTS: RISK FACTOR SCORE

Location	Population: Millions		Population: Fertility rate	Food and Agriculture: Land, % of potential arable land		Food and Agriculture: Inputs		Food and Agriculture: Food			Environment and Development: Fuel-wood situation (A) acute (D) deficit (PD) prospective deficit	Environment and Development: Income per caput 1980 $	Environment and Development: Agric. products on % of exports	Risk factor score
				cultiv.	irrig.	Kilos of fertilizer per hectare	Cereal yield tonnes per ha	Calorie supply as % of needs	Self-sufficiency ratio basic food					
	1975	2025	c 1975	1975	1975	1981	1981	1978-80	1978	2000			c 1980	
Africa														
Algeria	16	62.9	(7)	(89)	4	(32)	(0.74)	100	40	(35)	(D)	1940	1	6
Benin	3	13.9	(6.7)	51	0.1	(1.7)	(0.8)	100	95	(93)	(D)	(290)	(76)	7
Botswana	0.7	3.4	(6.5)	—	—	(1)	(0.33)	(94)			(A)	(880)	10	6/7
Burundi	3.8	3.4	(5.9)	(84)	0.5	(0.8)	(0.97)	(92)	99	(87)	(A)	(210)	(85)	9
Cape Verde	0.3	0.52	3	—	—	(2.5)	(0.3)	117	—	—	—	(310)	25	3/6
Comoros	0.3	1.02	(6.2)	—	—	—	(1.09)	(99)	—	—	(A)	(290)	(74)	6/6
Ethiopia	28.8	93.6	(6.7)	61	0.2	(4)	(0.86)	(74)	99	(68)	(A)	(130)	(94)	8
Kenya	13.5	82.3	(8)	60	0.6	(26)	1.72	(89)	101	(62)	(A)	(390)	(52)	7
Lesotho	1.2	3.7	(5.9)	—	—	(15)	(1.2)	107	—	—	(A)	(470)	(57)	6/7
Malawi	5.2	23	(6.9)	62	0.3	(14)	(1.4)	(96)	101	(67)	(D)	(190)	(85)	8
Mali	6.1	25	(6.7)	(87)	1.1	(6)	(0.69)	(85)	94	(59)	(A)	(180)	(90)	9
Mauritania	1.4	6.1	(6.9)	49	3	(11)	(0.52)	(89)	32	39	(A)	(400)	20	6
Mauritius	0.9	1.57	3.1	(95)	14	249	2.66	119	4	7	(A)	1120	(69)	3
Morocco	17.3	59.3	(6.6)	(91)	9	(33)	(0.91)	110	77	(66)	(D)	(830)	24	7
Namibia	0.9	3.3	(5.9)	—	—	—	(0.31)	(98)	—	—	(A)	1800	—	4/5
Niger	66	285	(6.9)	(76)	0.05	(6)	(0.74)	(99)	92	(87)	(D)	(870)	3	8
Nigeria	4.6	20.5	(7.1)	(96)	0.1	(0.8)	(0.35)	(94)	98	101	(A)	(300)	24	7
Reunion	0.5	0.82	3.1	—	—	206	5	128	—	—	(A)	3640	(88)	2/7
Rwanda	4.1	19.6	(6.9)	(91)	0.1	(0.1)	(1.13)	(95)	99	(85)	(A)	(220)	(87)	9
Senegal	5	16.8	(7.1)	54	2	(4)	(0.94)	100	55	(34)	(D)	(420)	34	6
Sierra Leone	3	10.7	(6.1)	(85)	0.1	(1)	(1.01)	(92)	92	(76)	PD	(300)	30	7
Somalia	3.1	13.4	(6.1)	55	2	(2)	(0.61)	(92)	59	(34)	(A)	(260)	(97)	8
Swaziland	0.5	1.9	(6.5)	—	—	71	(1.49)	108	—	—	(A)	(690)	(60)	5/7
Togo	21.3	8.9	(6.5)	(96)	0.15	(3)	(0.84)	(92)	95	(73)	(D)	(350)	36	8
Tunisia	5.6	13.1	(5.6)	(94)	2.6	(13.5)	(0.97)	115	61	72	(D)	1260	9	5

Uganda	11.3	51.9	(6.1)	60	0.1	(0.1)	(1.41)	(80)	100	(86)	(D)	(200)	(95)	8
Upper Volta	6.1	20.5	(6.5)	(83)	0.1	(4)	(0.57)	(85)	95	(75)	(D)	(200)	(79)	9
W. Sahara	0.12	0.36	—	—	—	—	(0.7)	—	—	—	—	—	—	1/1
Zimbabwe	6.2	28.4	(6.6)	14	0.3	65	1.72	(80)	102	(63)	PD	(780)	35	4
Asia														
Bangladesh	77	222	(6.2)	(99)	13	46	1.96	(81)	95	(77)	(D)	(130)	29	6
Bhutan	1.2	3.2	(6.1)	—	—	(1)	(1.08)	(41)	—	—	—	(80)	—	5/6
Philippines	43	108	4.8	(87)	12	(34)	1.6	102	96	101	(D)	(710)	32	4
Singapore	2.2	3.26	1.8	—	—	550	—	135	—	—	—	4420	8	0
Sri Lanka	13.6	26.8	3.6	(83)	17	77	2.43	101	71	90	(D)	(270)	(63)	4
Viet Nam	48	106	(5.2)	68	11	41	2.19	(94)	92	(71)	(D)	—	29	4/8
Southwest Asia														
Afghanistan	14	42.1	(6.9)	(89)	26	(6)	(1.27)	(75)	97	(57)	(A)	—	(45)	8/8
Bahrain	0.27	0.77	(7.6)	—	—	—	—	—	—	—	—	8200	1	1/3
Iran	32.7	99	(6.4)	(99)	32	36	(1.19)	121	75	(59)	—	—	1	4/8
Iraq	11	40.4	(6.9)	(97)	47	(17)	(0.99)	110	49	(33)	(D)	3020	0	6
Israel	3.5	7.5	3.5	—	—	199	2.53	118	—	—	—	4540	15	0/6
Jordan	2.7	11.2	(7.2)	(97)	6	(10)	(0.51)	(97)	13	18	(D)	1500	28	6
Kuwait	1	5	(6.1)	—	—	440	2.06	—	—	—	—	24160	0	1/5
Lebanon	2.8	5.6	4	(88)	19	76	2.15	101	18	(13)	PD	—	29	2/8
Oman	0.77	2.85	(7.1)	—	—	79	(1.33)	—	—	—	—	4840	1	2/5
Qatar	0.17	0.66	(6.8)	—	—	400	2.33	—	—	—	—	28600	0	1/5
Saudi Arabia	7.3	31	(7.2)	62	12	35	(0.67)	120	12	(10)	—	11950	0	3/8
Syria	7.4	32.7	(7.3)	(86)	8	(22)	(1.36)	115	89	(63)	(D)	1480	14	6
United Arab Emirates	0.5	2	(6.8)	—	—	269	2.4	—	—	—	—	28110	1	1/5
Yemen	5.3	16.3	(6.8)	(92)	13	(3.5)	(0.95)	(94)	69	71	(D)	(430)	(49)	8
Dem. Yemen	1.65	5.6	(6.9)	(89)	73	(10)	(1.54)	(86)	25	(23)	(D)	(420)	4	8
Central America														
Antigua	0.07	0.4	2.6	—	—	—	2.08	(91)	—	—	—	1380	9	1/5
Bahamas	0.2	0.45	3.5	—	—	75	(1.14)	(96)	—	—	—	3450	0	2/6
Barbados	0.24	0.38	2.1	—	—	176	2.5	126	—	—	—	3270	30	0/6
Dominican Republic	5.2	14.5	4.7	58	8	42	3.6	(94)	77	(67)	(D)	1190	(57)	4
El Salvador	4.1	15	(6.3)	(86)	9	89	1.74	(94)	85	(66)	(A)	(670)	(76)	7
Guadeloupe	0.3	0.4	2.7	—	—	73	(1.2)	113	—	—	—	3940	(89)	2/6
Guatemala	6.2	21.7	(5.7)	43	1.3	51	(1.51)	(94)	84	(78)	(D)	1080	(64)	6
Haiti	5.2	18.3	(5.5)	(85)	8	(0.4)	(0.82)	(83)	84	(55)	(A)	(280)	(62)	9
Jamaica	2.04	3.8	3.7	58	5	66	1.87	115	31	32	(A)	1090	15	1
Martinique	0.32	0.43	2.7	—	—	335	—	116	—	—	—	4340	38	0/5
Netherlands Antilles	0.24	0.48	—	—	—	—	(0.72)	108	—	—	—	4160	1	1/4
Puerto Rico	3.1	6.5	2.7	—	—	—	(0.74)	—	—	—	—	3220	—	1/3
Windward Islands	0.11	0.22	4.8	—	—	282	(0.73)	(99)	—	—	—	(890)	30	3/6
Trinidad & Tobago	1.08	1.8	2.7	(83)	9.5	51	2.77	112	15	21	—	5010	2	1/8

Notes on Table 1, Appendix 2

COLUMN

1. Population 1975 (Ref 20)
2. Population 2025 (Ref 20)
3. Fertility rate ca. 1975-80 (Ref 20)
4. Cultivated area 1974-76 as % of potential arable land (Ref 21)
5. Irrigated area as % of potential arable land (Ref 21)
6. Kilo/fertilizer/ha arable and permanent crops 1981 (Ref 7)
7. Cereal yield, t/ha, 1981 (Ref 10)
8. Calorie supply as % of requirements 1978-80 (Ref 19)
9. Self-sufficiency ratios in basic foods 1978-79 and 2000, trend (Ref 21)
10. Fuelwood situation: acute scarcity (A); deficit (D); potential deficit (PD) (Ref 15)
11. Income per caput 1980 (Ref 20)
12. Agricultural products as % of exports, ca. 1980 (Ref 19)
13. Risk factor score, based on following factors:
    1. Land scarcity (cultivating more than 70% of potential arable land, 1974-76)
    2. Fertilizer use below 35 kg/ha, 1981
    3. Cereal yields below developing market economy average (1.6 t/ha)
    4. Calorie supply below 100% of daily requirements, 1978-80
    5. Self-sufficiency in basic foods projected lower in 2000 than that in 1978-79
    6. Acute scarcity or deficit in fuelwood
    7. Fertility rate above 5
    8. Per caput income below US$1 000, 1980
    9. Agricultural products accounting for more than 40% of exports, ca. 1980

Note: Countries for which data are available on fewer than nine of these indicators are given the score out of nine that is the nearest whole number below their actual score expressed as a proportion of nine.

2. SUMMARY OF RISK FACTORS: 64 CRITICAL COUNTRIES (low inputs, 2000)

(N.B.: The figure in parentheses indicates the number of countries with a risk, the second, the number for which data were available.

Region Total no. of countries	Land scarce		Low fertilizer use		Low cereal yields		Calorie supply below require-ments		Self-sufficiency in basic foods declining		Fuelwood acute scarcity or deficit		Fertility rate above 5		Income below $ 1 600		Agricultural products more than 40% of exports	
	Higher Risk	Total	Higher Risk	Total	Higher Risk	Total	Higher Risk	Total	Higher Risk	Total	Higher Risk	Total	Higher Risk	Total	Higher Risk	Total	Higher Risk	Total
Africa (29)	(12)	21	(22)	26	(25)	29	(18)	28	(17)	21	(25)	27	(25)	28	(23)	28	(15)	27
SW Asia (15)	(8)	9	(7)	15	(9)	14	(4)	10	(7)	9	(6)	8	(13)	15	(2)	12	(2)	15
SE Asia (6)	(3)	4	(2)	6	(1)	5	(3)	6	(2)	4	(4)	5	(3)	6	(4)	5	(1)	5
Central America (14)	(3)	6	(1)	11	(7)	13	(7)	13	(4)	6	(5)	5	(3)	13	(3)	14	(5)	13
Total (64)	(26)	40	(32)	58	(42)	61	(32)	57	(30)	40	(40)	45	(44)	62	(32)	59	(23)	60
% of critical countries with higher risk	65%		55%		69%		56%		75%		89%		71%		54%		38%	

APPENDIX 2.3

THE MOST CRITICAL COUNTRIES AND TERRITORIES, YEAR 2000 (LOW INPUTS)

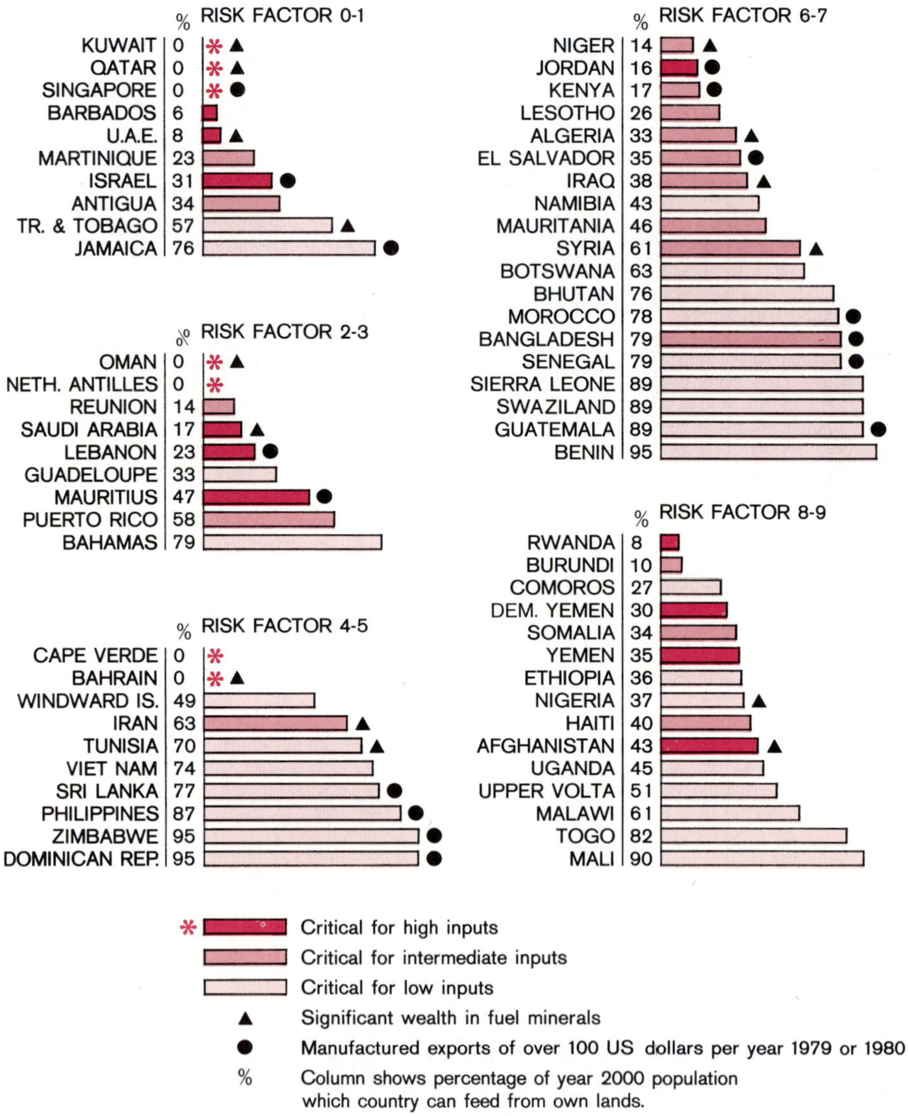

Appendix 3

Technical note

Assessing input levels by yield

Assessing input levels by yield

The assessment was done as follows: for each crop, the average 1979-81 yield in developing market economies in Africa was taken to equate with the low level of inputs, since African yields approximated medium results for the low-input level. The length-of-growing-period zone which gave the closest yield to that average was selected.

The yields corresponding to intermediate and high levels of inputs could then be read off (Ref. 9).

The yields obtained by the other regions for their major food crops were then compared with these guidelines. The rating in the table below is based on taking the difference between the low-input figure and the intermediate-input figure as 100 percent. The rating was done for wheat in the Near East, for rice and wheat in the Far East, and for maize and wheat in Latin America. For the latter two regions, the rating in the chart represents the average for the two crops, weighted to account for the area under each one (Ref. 10).

Guideline yields (tonnes/hectare)

Crop	Low	Level of inputs: Intermediate	Level of inputs: High	For comparison (W. Europe)
Wheat	0.9	2.3	3.7	(3.69)
Maize	1.2	3.5	5.4	(5.08)
Rice	1.3	3.25	5.07	(5.47)

Average yields achieved (1979-81) and area

	Near East	Far East	Latin America
Wheat	1.46	1.55 (30 million ha)	1.49 (9.7 million ha)
Maize	—	—	1.82 (27.5 million ha)
Rice	—	2.19 (84 million ha)	—